NUMERICAL METHODS FOR ENGINEERS

A Practical Approach

NUMERICAL METHODS FOR ENGINEERS
A Practical Approach

Abdulmajeed A Mohamad

University of Calgary, Canada

Adel M Benselama

College of Mechanical and Aerotechnical Engineering,
ISAE–ENSMA, Poitiers, France

World Scientific

W JERSEY · LONDON · SINGAPORE · BEIJING · SHANGHAI · HONG KONG · TAIPEI · CHENNAI · TOKYO

Published by

World Scientific Publishing Co. Pte. Ltd.

5 Toh Tuck Link, Singapore 596224

USA office: 27 Warren Street, Suite 401-402, Hackensack, NJ 07601

UK office: 57 Shelton Street, Covent Garden, London WC2H 9HE

Library of Congress Cataloging-in-Publication Data

Names: Mohamad, A. A. (Abdulmajeed A.), author. | Benselama, Adel, author.

Title: Numerical methods for engineers : a practical approach /

    Abdulmajeed A Mohamad, University of Calgary, Canada,

    Adel Benselama, College of Mechanical and Aerotechnical Engineering, France.

Description: Singapore ; Hackensack, NJ ; London : World Scientific, [2023] |

    Includes bibliographical references and index.

Identifiers: LCCN 2022025574 | ISBN 9789811255250 (hardcover) |

    ISBN 9789811255267 (ebook for institutions) | ISBN 9789811255274 (ebook for individuals)

Subjects: LCSH: Engineering mathematics. | Numerical analysis.

Classification: LCC TA335 .M64 2023 | DDC 620.001/51--dc23/eng/20220701

LC record available at https://lccn.loc.gov/2022025574

British Library Cataloguing-in-Publication Data

A catalogue record for this book is available from the British Library.

For any available supplementary material, please visit
https://www.worldscientific.com/worldscibooks/10.1142/12809#t=suppl

Desk Editors: Aanand Jayaraman/Steven Patt

Typeset by Stallion Press
Email: enquiries@stallionpress.com

The first author (A.A. Mohamad): To my lovely wife Lubna Younis

The second author (A.M. Benselama): To my parents, children, and my beloved spouse Nivine, who passed away, a role model of a lionhearted lady

Preface

After teaching numerical and computational courses for undergraduate and graduate students, we realized that the basics of the course are not plentiful and most of the applications can be worked out by knowing those fundamentals. The Taylor series and linear and nonlinear algebra are the keys to most of the numerical methods. The Taylor series assumes that the function is continuous. The outcome of differential equations is a system of algebraic equations. This book focuses on introducing the essential elements or the basis of numerical methods and linking the applications to those crucial elements, i.e., Taylor series and algebraic equations (linear or nonlinear equations). With such an approach, students can realize that numerical methods are based on a few fundamentals. We also tried to eliminate less popular methods and focused on the methods that work for most problems. Such a coherent approach eliminates confusion about whether a method works or not for a specific problem. Numerical techniques are not only a science but also an art. Hence, practicing more problems helps in selecting the proper scheme and getting around issues that causes problems — "learning the tricks of the trade."

One main question arises these days with the advances in software, such as Mathematica, Maple, and CFD codes: Do we need to teach students how to develop numerical codes or teach them how to use the available codes? The same question may be put in another way: There are many automobiles in the market, do we need to make another automobile? Also, we may argue why we need to teach calculus or other mathematical methods, while Mathematica or Maple can be used more efficiently, and so on. The students should know, what is embedded in those closed boxes (software). For many reasons, a better understanding of the method's limitations is essential in obtaining the correct results. Extending or developing methods for specific applications are needed, where the available codes or software are difficult to modify or costly.

Moreover, by mastering the numerical techniques, students who may use them in their career to solve specific problems, need not rely on an expensive software. In nature, most phenomena are controlled by a few physical laws. We have to show the whole picture as a unified field rather than drawing artificial borders between the topics. We are living in a connected and globalized era. Then, why don't we think the same way in our teaching? In this textbook, we tried our best to build a coherent approach as much as possible. Our main aim in writing this text is that it should reflect our many years of experience and to share our accumulated experience with the readers. However, we are not claiming that we did a faultless job — "No perfect man-made book has been published yet." We are open to any suggestions, which will be highly appreciated and will be acknowledged in future editions. Your comments can be communicated via email: mohamad@ucalgary.ca or adel.benselama@ensma.fr.

About the Authors

Dr. Abdulmajeed A. Mohamad is Professor, Department of Mechanical and Manufacturing Engineering, The University of Calgary, Canada. He is also Interim Director of Centre for Environmental Engineering for Research and Education (CEERE). Dr. Mohamad is an ASME fellow member and Executive Scientific member of the International Center for Heat and Mass Transfer (ICHMT). Dr. Mohamad's research area spans across different aspects of thermal sciences and engineering. He is extensively contributed to the fundamentals of heat and mass transfer in natural convection, combustion in porous media, flow and heat transfer in liquid metals, solar energy, adsorption/desorption, thermal radiation, etc. His work on solar energy resulted in the development of a high-efficiency solar air heater. Also, his work on combustion in the porous medium resulted in an ultra-low NOx burner. Dr. Mohamad published over 300 papers, mainly in peer-reviewed journals.

Dr. Mohamad extensively contributed to the computational fluid flow, heat, and mass transfer. He is the author of a popular textbook on *Lattice Boltzmann Method, Fundamentals and Engineering Applications with Computer Code.* The second edition of the book was published by Springer 2019. Also, the text was translated to the Chines language by Springer.

Dr. Mohamad is the founder of the International Conference on Computational Heat and Mass Transfer (ICCHMT) and Applications on Porous Media. The 13th edition of ICCHMT was held in Paris, May 2021.

Dr. Mohamad has been invited by many universities and institutes worldwide as a visiting professor, lecturer, keynote speaker, Ph.D. thesis examiner, program evaluator, collaborator.

While he was on special leave from The University of Calgary, he was acting Dean of Engineering at Alfaisal University, Saudi Arabia. He was instrumental in establishing an engineering school there.

In recognition of his contribution, *International Journal of Heat and Mass Transfer* has published an article on his 65 birthday.

Assistant Professor Adel M. Benselama received his M.Eng. in Aeronautics at National Institute of Aeronautics of Blida, Algeria, his M.Sci. and Ph.D. in fluid mechanics, heat and mass transfer, from University of Grenoble, France. He had three Postdoctoral fellowships at INP of Grenoble and at University Hauts-de-France. Dr. Benselama joined the Higher National College of Mechanical and Aerotechnical Engineering, ISAE-ENSMA, France, in Fall 2011. His is currently teaching numerical methods, statistical physics and heat transfer. Dr. Benselama field of research is transport phenomena involving moving interfaces, convection, conduction, radiation and combinations of. He has published over than 40 papers in the lead journals in the field. He has made contributions in the area of interface flows instabilities due to capillarity and thermocapillarity, electrohydrodynamics, microscale contact line physics, low Reynolds number flows, high Mach number flows and also radiation in semi-transparent media. He has also worked extensively in advanced numerical methods involving finite element, finite volume and boundary element methods.

Acknowledgment

We think that "No job can be done by an individual." Any job we did was based on or built on the knowledge we learned from others and questions imposed by others, which helps us think and learn. We don't want to use war an example, but it may be appropriate to say that many unknown soldiers die on the battlefield to ensure the livelihood of others on peaceful lands. We learned a lot from the people we worked with and the students we taught. But without sound and relaxed environments, nothing can be done as it is supposed to be. Smiles and words of encouragement from the people around you give more energy to work than any material support. Thanks to all our former students. While we were teaching them, we learned from them.

The second author would like to acknowledge the help of his colleagues Mr. Goncalves Da Silva, E., and Mr. Virot, F., among others, for their useful material.

Contents

Preface   vii

About the Authors   ix

Acknowledgment   xi

1.   Fundamentals: Taylor Series   1

   1.1   Approaches to Solve a Problem   1
       1.1.1   Experimental   1
       1.1.2   Analytical and computational
              approaches   2
       1.1.3   Computational or numerical
              approaches   2
   1.2   Mathematical Modeling   3
       1.2.1   Divide and rule   3
       1.2.2   Non-dimensionalization   5
   1.3   Code Development Procedure   6
   1.4   Errors Associated with Modeling and
       Computational Methods   7
       1.4.1   Round-off errors   7
       1.4.2   Truncation errors   8
       1.4.3   Discretization errors   8
   1.5   Taylor Series   8
       1.5.1   One-variable functions   8
       1.5.2   Multi-variable functions 13
   1.6   Problems 15

2.   Linear Algebra   17

   2.1   Linear Algebra   17
   2.2   Vectors and Matrices   17
       2.2.1   Definitions and properties   17
       2.2.2   Examples   24
   2.3   System of Linear Algebraic Equations   26
   2.4   Direct Methods   26
       2.4.1   Gauss elimination method   27
       2.4.2   Thomas algorithm   29
       2.4.3   Gauss–Jordan method   31
       2.4.4   LU factorization (decomposition)   31
   2.5   Symmetric Matrix   33
   2.6   Iterative Methods   34
   2.7   Stationary Iterative Methods   34
       2.7.1   Jacobi method   37
       2.7.2   Gauss–Seidel method   38
       2.7.3   Examples using Jacobi and Gauss–Seidel
              methods   39
       2.7.4   Successive over-relaxation method   40
   2.8   Non-stationary Methods   43
       2.8.1   Steepest descent method   43
       2.8.2   Conjugate gradient method   48
   2.9   Problems   51
   Extra Reading   52

3.   Interpolation and Fitting   53

   3.1   Background   53
   3.2   Linear Interpolation   54
   3.3   Higher-Order Interpolation   55
   3.4   Lagrange Interpolation   55
   3.5   Chebyshev Polynomial   58
   3.6   Least Square Method   64
       3.6.1   One-variable fitting functions   64
       3.6.2   Multi-variable fitting functions   65
   3.7   Spline Interpolation   67

      3.7.1   Quadratic splines 67
          3.7.1.1   Basic quadratic splines 67
          3.7.1.2   Generalized quadratic splines . . . 69
      3.7.2   Cubic splines 70
  3.8   Problems 72

4. Nonlinear Equations   **75**

  4.1   Bisection Method 76
  4.2   Fixed Point Method 77
  4.3   Newton–Raphson Method 79
  4.4   Chebyshev Method 81
  4.5   Secant Method 82
  4.6   Problems 82
  4.7   System of Nonlinear Algebraic Equations 85
  4.8   Problems 87

5. Numerical Differentiation and Integration   **91**

  5.1   Introduction to Numerical Differentiation 91
  5.2   First Derivative 92
      5.2.1   Forward, backward, and central first-order
             approximations 92
      5.2.2   Higher-order approximations for first
             derivative 95
  5.3   Higher-Order Derivatives 95
  5.4   Undetermined Coefficient Method 97
  5.5   Summary of Most Used Derivatives 100
  5.6   Non-uniform Grids 102
      5.6.1   First derivative 103
      5.6.2   Second derivative 104
  5.7   Introduction to Numerical Integration 106
  5.8   Gauss Quadrature 109
      5.8.1   Increased flexibility 110
  5.9   Monte Carlo Method 112
  5.10   Error and Extrapolation 113
  5.11   Problems 114

6.   **Ordinary Differential Equations: Initial Value Problems**   117

   6.1   Introduction 117
   6.2   Initial Value Problems 118
   6.3   General Form of Initial Value, First-Order
          Ordinary Differential Equations 119
   6.4   Semi-analytical Method 120
   6.5   Taylor Series Method 121
          6.5.1   Euler method 122
          6.5.2   Explicit Euler method 123
          6.5.3   Implicit Euler method 124
          6.5.4   Central scheme 124
   6.6   Higher-Order Methods 126
          6.6.1   Taylor series method 126
          6.6.2   Fourth-order Runge–Kutta method 127
          6.6.3   Polynomial methods 129
          6.6.4   Verlet method 130
   6.7   Nonlinear ODE 131
   6.8   Order of Accuracy, Consistency, Stability,
          and Convergence 133
          6.8.1   Order of accuracy and consistency 133
          6.8.2   Consistency analysis 133
          6.8.3   Stability analysis 135
          6.8.4   Convergence 136
   6.9   Examples 137
   6.10   Problems 139
   Extra Reading 144

7.   **Ordinary Differential Equations: Boundary Value Problems**   145

   7.1   Introduction 145
   7.2   Linear Shooting Method 146
   7.3   Nonlinear Shooting Method 147
          7.3.1   Secant method 148
          7.3.2   Newton–Raphson method 148
   7.4   Finite Difference Method 150
          7.4.1   Planar problems 150
          7.4.2   Axisymmetric problems 150

7.5 Boundary Conditions 151

    7.5.1 Modified equation 153

    7.5.2 Example 154

7.6 Finite Volume Method 154

    7.6.1 Physics approach 155

    7.6.2 Local governing equation approach 159

    7.6.3 Example: Finite difference and shooting method 160

    7.6.4 Example: Finite volume method 161

7.7 Problems 166

Extra Reading 168

8. Partial Differential Equations   **169**

8.1 Introduction 169

8.2 Appropriateness of a Numerical Method 170

    8.2.1 Lax–Richtmyer theorem 170

    8.2.2 Example of existence and uniqueness of the solution 172

8.3 Consistency Analysis 172

8.4 Stability Analysis 174

    8.4.1 A first example 176

    8.4.2 A second example 177

9. Diffusion Equation (Parabolic Equation)   **179**

9.1 Introduction 179

9.2 Finite Difference 180

    9.2.1 Explicit methods 180

    9.2.2 Implicit methods 182

    9.2.3 Crank–Nicolson method 183

9.3 Unstable and Inconsistent Schemes 183

9.4 Multi-dimension 186

9.5 Alternating Direction Implicit Schemes 187

9.6 Finite Volume Method 189

    9.6.1 The key idea 189

    9.6.2 Boundary conditions for FVM 192

9.7 Worked Example 193

10. Laplace and Poisson Equations (Elliptic Equations)   205

   10.1 Introduction 205
   10.2 Finite Difference Method 206
         10.2.1 Worked example 206
   10.3 Finite Volume Method 214
   10.4 Methods of Solution 215
   10.5 Problems 217

11. Advection and Advection–Diffusion Equations   221

   11.1 Explicit Methods 221
         11.1.1 Lax–Wendroff method 222
         11.1.2 Upwind scheme 224
         11.1.3 Lax method 224
         11.1.4 Leapfrog method 225
   11.2 Implicit Methods 225
         11.2.1 Implicit Euler method 225
         11.2.2 Crank–Nicolson method 226
   11.3 Multi-step Methods 226
         11.3.1 Richtmyer/Lax–Wendroff multi-step
                method 226
         11.3.2 MacCormack method 227
   11.4 Nonlinear Problems 228
         11.4.1 Explicit methods 228
         11.4.2 Implicit first-order upwind scheme 229
   11.5 Runge–Kutta Method 229
   11.6 Advection–Diffusion Problems 230
         11.6.1 Central difference in space scheme 230
         11.6.2 MacCormack scheme 231
   11.7 Finite Volume Method Applied to Advection
        Equation 231
         11.7.1 Centered schemes 233
                11.7.1.1 Basic formulation 233
                11.7.1.2 Jameson–Schmidt–Turkel
                        scheme 234
         11.7.2 Flux vector splitting schemes 235
                11.7.2.1 Introduction with a scalar
                        equation 235

11.7.2.2 Extension to a hyperbolic system of equations 236

11.7.2.3 Modified Steger and Warming scheme 236

11.7.2.4 Vijayasundaram scheme 237

11.7.2.5 van Leer scheme 237

11.7.3 Riemann solvers: Flux difference splitting 238

11.7.3.1 Godunov's scheme 238

11.7.3.2 Roe's scheme 239

11.7.3.3 Harten, Lax, and van Leer schemes 240

11.7.3.3.1 Integral formulation of Riemann's problem: 241

11.7.3.3.2 Harten, Lax, and van Leer basic scheme 242

11.7.3.3.3 Harten, Lax, and van Leer contact scheme 243

11.7.4 Liou and Steffen AUSM scheme and its variant 245

11.7.4.1 Liou's basic AUS method 245

11.7.4.2 Liou's AUSM+ 247

11.7.5 Higher-order schemes 247

11.7.5.1 Monotonic upstream-centered scheme for conservation law . . . 248

11.7.5.2 TVD schemes and limiters 250

11.7.6 A worked example 253

11.8 Finite Volume Method Applied to Advection–Diffusion Equation 254

11.9 Examples 259

11.10 Vorticity Stream Function Formulation 261

Extra Reading 264

12.   Wave Equation   265

  12.1   Consistency 267
  12.2   Stability Analysis 268
  12.3   Worked Example 269
  12.4   Lax–Wendroff Method 271
  12.5   Implicit Method 272
  12.6   Problems 273

Index   277

Chapter 1

Fundamentals: Taylor Series

In this chapter, the fundamentals of modeling are discussed, with more attention given to Taylor series. Taylor series is one of the main pillars of many numerical techniques, as will be shown in the following chapters. A continuous function can be expanded using Taylor series. If our interest is to evaluate a function at a point not that far from a known point, the function can be approximated or simplified and even linearized at that point. The beauty and applicability of Taylor series will be explored after a few chapters. In this chapter, Taylor series is constructed and its properties are discussed. However, before we dive into numerical techniques, let us discuss a few items related to mathematical modeling and solution methodologies.

1.1 Approaches to Solve a Problem

In general, there are three approaches used by scientists and engineers to solve a problem, namely experimental, analytical, and computational. Each of these approaches has advantages and disadvantages.

1.1.1 Experimental

The experimental approach involves working on real, practical physical problems. It has to do with sensing the problem and observing the phenomenon. The experimental approach is used to validate simulation results and hypotheses. However, for certain problems, it is impossible to perform experiments. More often, it is difficult to control and isolate the effects of parameters. Also, experimental errors are unavoidable. In most cases, performing experiments is costly and time-consuming. However, experiments are essential for any application. In most cases, after analysis is performed either theoretically or computationally, experiments must be carried out to validate the predicted results, especially for complex problems and/or for new problems.

1.1.2   Analytical and computational approaches

The elegance of the analytical approach lies in closed-form equations and in the fact that the effect of each controlling parameter can be examined and evaluated. However, analytical approaches are limited to simple and, in most cases, linear problems. Also, handling complex geometries is difficult analytically. Fortunately or unfortunately, most engineering and scientific problems are nonlinear and/or involve complex geometries, which are difficult to solve analytically. Therefore, numerical methods come into the picture as a cost-effective approach that can be adopted in a relaxed environment. The outcome of most numerical simulations is massive data, which requires data reduction and visualization. The predicted data can be in the order of mega- to gigabytes. However, advances in computer visualizations and storage systems have brought simulations to our desktop machines.

1.1.3   Computational or numerical approaches

In general, it is possible to handle a wide range of problems, linear and nonlinear, with complex geometries, numerically. It is possible to isolate the effects of parameters. The algorithm of the numerical approach is as follows:

- Identify the problem and the objective of the analysis.
- Identify the equations that govern the physics of the problem.
- Specify the boundary and initial conditions.
- Most practical problems are unclear or difficult to handle at the first stage of the analysis; different assumptions are needed to be made to simplify the problem. Those assumptions can be relaxed later on.

- Sketch the problem using a coordinate system, if needed.
- Treat the problem numerically (there are many numerical methods available, such as finite difference, finite element, finite volume, Lattice–Boltzmann method, spectral method, boundary element method, and molecular dynamics). The outcome of the numerical approximation is a set of linear or nonlinear algebraic equations (order of thousands, millions, or beyond). Solve those equations numerically, mainly iteratively.
- Analyze the output data and ensure that the data do not violate the physics of the problem. Compare the data with some available experimental data and/or with published numerical predictions. This process is called validation.
- Also, it is necessary to ensure that the predicted results are independent of the grid size and time step (for unsteady-state problems). This process is called verification.

1.2   Mathematical Modeling

Mathematical modeling is a systematic methodology of converting a physical problem into a mathematical form (mainly equations). Let us assume that we have a problem, such as a new aircraft design, air-conditioning a building, understanding mass transfer exchange in the human lung, underground water movement, CO_2 capture, very long bridge construction, heat exchangers, boilers, turbines, microfluidics, blood flow, etc. These problems are three-dimensional, time dependent, nonlinear, involve complex geometry, exposed to a dynamical environment, etc.

1.2.1   Divide and rule

The first step in modeling is to simplify the problem, i.e., instead of starting with a three-dimensional problem, assume one- or two-dimensionality and neglect the effect of less-significant dimensions and less-important parameters. Consider representative but minimalistic (filtered) objects of scrutiny, see Fig. 1.1, assume that the system is exposed to static conditions, etc. By following these steps, we are deviating from the actual problem, the impact of which depends on the nature of that problem and on our physical understanding of it.

Figure 1.1: From complicated object (left) to a tractable simplified model (right). (The object in question is a steam locomotive of the mid-nineteenth century. Artwork by Freepik.)

More complicated and sophisticated models can be considered at a later stage.

Also, the initial and boundary conditions we impose on the model are never exact. Hence, another type of error due to approximating the boundary and initial conditions arises. In most cases, we need to identify some input material properties, such as viscosity, thermal conductivity, and modulus of elasticity. Again, in most cases, we do not have exact values for those properties. Also, the properties may change with the problem conditions, such as temperature and stress. The effects of those approximations and errors associated with the estimations are difficult to generalize because they are problem dependent.

The second step is to identify the governing equations. Those equations, in general, are not exact but reflect the "main" physics of the problem. Commonly, the outcome of the mathematical modeling is a set of algebraic, ordinary or partial differential equations. Most problems are continuous, for which the analytical solution is difficult due to the nonlinearity of the problem and/or complexity of the geometry. Hence, numerical methods are our avenue to solve those equations. The first step of the numerical (digital) solution is to convert those differential equations (continuous) into a set of algebraic equations (discretized) using a method, such as finite difference, finite volume, or finite element. Selecting a method depends on its relevance and on our background. The outcome of the discretization is a system of algebraic equations (linear or nonlinear). Therefore, instead of trying to solve differential equations, we end up solving algebraic equations. However, the price we pay is more errors introduced to the process due to **truncation** procedure. Also, the domain of interest needs to be discretized into grids, finite volumes, finite elements, etc. Again, another type of error is introduced, called **discretization errors**. Solving those algebraic equations is not exact because of **round-off errors**. The optimal numerical method solver is the one that minimizes numerically introduced errors (from digitizing or discretizing the continuous domain to the final solution of the algebraic equations).

However, before solving the governing differential equations numerically, it is preferable to scale those equations or make them dimensionless. Such a process is essential in data reduction and generalization of the predicted results. In the following section, we discuss this process with examples.

1.2.2   Non-dimensionalization

Most physical systems have basic units, e.g., length, time, mass, and temperature. Hence, the dependent and independent variables have basic or a combination of basic units. The model equations may also contain physical properties, such as density and viscosity. It is a good strategy to manipulate the governing equations and associated boundary and initial conditions to formulate them in non-dimensional (unitless) terms. Such a process has a great advantage in that the resulting outcome of the simulation will be general. To illustrate the process, let us work on an example.

Example: Initially, a stainless steel slab of 2 m thickness is at room temperature (20 °C). The left-hand surface is subjected to a high temperature of 200 °C. The right-hand surface is kept at the room temperature. Calculate the temperature distribution in the slab at different time intervals.

Assuming constant thermophysical properties and adiabatic lateral walls, the governing equation, boundary conditions, and initial condition for the problem are

$$\frac{\partial T}{\partial t} = \alpha \frac{\partial^2 T}{\partial x^2}, \tag{1.1}$$

at $x = 0$, $T = 200$ °C and at $x = L = 2$ m, $T = 20$ °C, and at time $(t) = 0$, $T = 20$ °C for all x, respectively.

We can solve the above problem as it is, and our results will be valid only for the specific conditions as stated above. However, we can define non-dimensional variables, namely $\theta = \frac{T-T_0}{T_h-T_0}$ and $\eta = \frac{x}{L}$, where T_0, T_h, and L are room temperature, left-hand surface temperature (hot surface), and length of the slab, respectively. Also, let us define dimensionless time (τ) as $\tau = \frac{\alpha t}{L^2}$. The governing equations and associated boundary and initial conditions can be written as follows:

$$\frac{\partial \theta}{\partial \tau} = \frac{\partial^2 \theta}{\partial \eta^2}, \tag{1.2}$$

at $\eta = 0$, $\theta = 1$ and at $\eta = 1$, $\theta = 0$, and at time $\tau = 0$, $\theta = 0$.

The outcome of the solution will be general for any length of slab, for any boundary and initial temperatures. Moreover, the results are applicable for any kind of slab material.

Note: Selection of the reference parameters depends on the problem. However, those reference parameters should be constant and should not change with the solution.

1.3   Code Development Procedure

Developing a simulation code may take a long time. However, the debugging process may take even more time. It is more manageable to develop a code in modules. In other words, a large code (hundreds of lines) can be developed using subroutines and functions. This strategy facilitates two things. First, each subroutine or function can be tested before being added to the main program. Second, it becomes easy to modify the code to replace subroutines or functions.

It is essential to develop a code by using a step-by-step procedure. It is highly recommended to start with a simple code, where experimental and/or analytical data are available. Then, relax the assumptions step by step. For example, if the problem involves variable properties, the developed code should be first tested for a constant-properties problem. Another example: Consider a flow in a duct filled with porous medium. The code should first be tested for a flow in a hollow duct. The fully developed flow can be checked with the analytical solution. The code's prediction should reflect the physics of the problem. For instance, by decreasing the permeability (Darcy number), the code should predict a flatter velocity profile.

1.4 Errors Associated with Modeling and Computational Methods

In general, there are two types of errors associated with modeling and numerical simulations. The errors associated with modeling may arise from selecting incorrect properties, assuming that the properties are constant, etc. This kind of errors is input error. An example is assuming incorrect physics, such as adding an extra source term to the momentum or to the energy equations and studying the effects of those terms. However, practically such a phenomenon does not exist. An example is adding a magnetic effect to airflow, when the air is not a magnetic material. Of course, the results show those effects because of the artificial source term. Therefore, validation of the numerically predicted data against experimental data becomes necessary. This does not mean that the experimental data are exact. However, the validation provides confidence in the predicted results. The problem is that not always we have experimental data for a given problem. **Therefore, understanding the physics of the problem is very important to get a qualitative sense of the results**. The above-mentioned type of errors are problem dependent and in no way related to the numerical solutions of the problem. The errors associated with selecting proper numerical schemes and method of solution come under the topic of numerical methods. Analyzing these type of errors is called verification. A list of errors associated with numerical methods are as follows.

1.4.1 Round-off errors

The number of decimal digits that can be carried by a computer is limited, e.g., 32, 64, and 128. Since numerical methods deal with arithmetic manipulations of the digits, million or even billion times, those errors accumulate. Hence, increasing the number of equations to solve has advantages and disadvantages. We will see that it is highly recommended to increase the number of nodes or meshes to get accurate solutions. However, at the same time, as the number of equations (number of arithmetic manipulations) increases, the round-off error increases too. In most problems, the round-off error does not cause a problem. However, for ill-conditioned problems, it does. A partial remedy is to use double precision or even quadruple precision, but it is definitely not a universal panacea.

1.4.2   Truncation errors

In most numerical analyses, differential equations are approximated using series expansions (Taylor or Fourier series), which has to be truncated at a certain term. Consequently, the approximation introduces an error which depends on the order of the approximation. Low-order approximation leads to wrong physics because the remainder of the approximation adds extra hidden terms to the solution (called numerical diffusion or numerical dispersion, depending on the order of approximation). Keeping more terms increases the round-off error and complicates the algorithm. Also, in some cases, it leads to an unstable solution unless a proper action is taken. Then, there will be a trade-off.

1.4.3   Discretization errors

In numerical analyses, not only are the equations approximated but also the continuous domain of integration discritized to elements, nodes, control volumes, etc., which also adds another type of error.

1.5   Taylor Series

1.5.1   One-variable functions

Taylor series can be constructed for any continuous function. For simplicity, let us consider a one-variable function, $f(x)$. Hence, the zero-order function can be represented as

$$f(x) = c_0, \tag{1.3}$$

where c_0 is a constant at a given value of x. The above function is a horizontal line parallel to x-axis at a distance $|c_0|$ from the x-axis. Let us evaluate the function at a given value, $x = a$, hence $f(a) = c_0$.

If we tilted the line by an angle, then $f(x)$ represents a straight line, i.e.,

$$f(x) = c_0 + c_1(x - a). \qquad (1.4)$$

An expression for the constant c_1 can be obtained by differentiation of the above equation, i.e.,

$$\frac{df(x)}{dx} = c_1. \qquad (1.5)$$

Since we know the value of the function and its derivative at $x = a$, then Eq. (1.4) can be rewritten as

$$f(x) = f(a) + \frac{df(a)}{dx}(x - a). \qquad (1.6)$$

Furthermore, if we add a quadratic term (or curvature) to Eq. (1.4) as

$$f(x) = c_0 + c_1(x - a) + c_2(x - a)^2, \qquad (1.7)$$

then by taking the second derivative of the function $f(x)$ with respect x, Eq. (1.7) yields

$$\frac{1}{2}\frac{d^2 f}{dx^2} = c_2. \qquad (1.8)$$

By substituting the first and second derivatives evaluated at $x = a$, Eq. (1.7) can be rewritten as

$$f(x) = f(a) + \frac{df(a)}{dx}(x - a) + \frac{1}{2}\frac{d^2 f(a)}{dx^2}(x - a)^2. \qquad (1.9)$$

We can repeat the procedure of adding more and more terms. We end up with a series called Taylor series:

$$f(x) = f(a) + \frac{df(a)}{dx}(x - a) + \frac{1}{2!}\frac{d^2 f(a)}{dx^2}(x - a)^2$$
$$+ \frac{1}{3!}\frac{d^3 f(a)}{dx^3}(x - a)^3 + \cdots + \frac{1}{n!}\frac{d^n f(a)}{dx^n}(x - a)^n. \qquad (1.10)$$

Let $\Delta x = (x - a)$, then the above equation can be written as

$$f(x) = f(a) + \frac{df(a)}{dx}\Delta x + \frac{1}{2!}\frac{d^2 f(a)}{dx^2}\Delta x^2$$
$$+ \frac{1}{3!}\frac{d^3 f(a)}{dx^3}\Delta x^3 + \cdots + \frac{1}{n!}\frac{d^n f(a)}{dx^n}\Delta x^n. \qquad (1.11)$$

If $a = 0$, Eq. (1.10) will become

$$f(x) = f(0) + \frac{df(0)}{dx} + \frac{1}{2!}\frac{d^2 f(0)}{dx^2}x^2 + \frac{1}{3!}\frac{d^3 f(0)}{dx^3}x^3 + \cdots + \frac{1}{n!}\frac{d^n f(0)}{dx^n}.$$
$$(1.12)$$

The above equations represent **forward expansion** if point x is on the right-hand side of point a ($x > a$). They represent backward expansion if the point x is on the left-hand side of point a ($x < a$). For backward expansion, Eq. (1.10) can be rewritten as

$$f(x) = f(a) - \frac{df(a)}{dx}(x - a) + \frac{1}{2!}\frac{d^2 f(a)}{dx^2}(x - a)^2$$

$$- \frac{1}{3!}\frac{d^3 f(a)}{dx^3}(x - a)^3 + \cdots + \frac{(-1)^n}{n!}\frac{d^n f(a)}{dx^n}(a - x)^n, \quad (1.13)$$

i.e.,

$$f(x) = f(a) - \frac{df(a)}{dx}\Delta x + \frac{1}{2!}\frac{d^2 f(a)}{dx^2}\Delta x^2$$

$$- \frac{1}{3!}\frac{d^3 f(a)}{dx^3}\Delta x^3 + \cdots + \frac{(-1)^n}{n!}\frac{d^n f(a)}{dx^n}\Delta x^n. \quad (1.14)$$

Example 1: Approximate function $y = 2 + 3x^3$ using Taylor series near $x = 0$.

Solution: First, we need to find the derivatives of the function and evaluate them at $x = 0$:

$$\frac{dy}{dx} = y' = 9x^2 \qquad y'(0) = 0$$
$$y'' = 18x \qquad y''(0) = 0$$
$$y''' = 18 \qquad y'''(0) = 18$$

Hence, $y = 2$ if only the first term is considered (zero approximation), $y = 2$ if two terms are considered (first-order approximation), $y = 2$ if three terms are considered (second-order approximation), and $y = 2 + 0 + 0 + \frac{1}{3!}18x^3$, which is equal to $y = 2 + 3x^3$ if four terms are considered (third-order approximation).

Example 2: Approximate $f(x) = \sin x$ using Taylor series. Calculate the value of $\sin(0.1)$ and $\sin(0.7)$.

Solution: Since we know the value of the function at $x = 0$, $\sin(0) = 0$, let us expand the function at $x = 0$:

$$\frac{df(x)}{dx} = \cos x \qquad f'(0) = 1$$
$$f'' = -\sin x \qquad f''(0) = 0$$
$$f''' = -\cos x \qquad f'''(0) = -1$$
$$f'''' = \sin x \qquad f''''(0) = 0$$

$$\vdots \qquad\qquad \vdots$$

Then, $\sin x$ can be approximated as

$$\sin x = 0 + x - 0 - \frac{x^3}{3!} + 0 + \frac{x^5}{5!} - 0 - \frac{x^7}{7!} + \cdots \tag{1.15}$$

or

$$\sin x = x - \frac{x^3}{3!} + \frac{x^5}{5!} - \frac{x^7}{7!} + \cdots + (-1)^{n+1}\frac{x^{2n-1}}{(2n-1)!} \quad n = 1, 2, \ldots, \tag{1.16}$$

$$\sin(0.1) = 0.1 - 0.1^3/6 + 0.1^5/120 - \cdots = 0.0999834\ldots,$$
$$\sin(0.7) = 0.7 - 0.7^3/6 + 0.7^5/120 - \cdots = 0.644233916\ldots.$$
Use your hand calculator to evaluate the above results. You will find that as the x value increases, you need more terms to evaluate $\sin x$. In fact, your calculator performs series evaluation each time you hit sine or cosine key.

Example 3: Approximate $f(x) = \sin x$ using Taylor series up to polynomials of 1st, 7th, 9th, and 17th degrees and compare the results on a graph for the interval $[-4\pi, 4\pi]$.

Solution:

$$f^{(0)} = \sin x \qquad f^{(0)}(0) = 0$$
$$f^{(1)} = \cos x \qquad f^{(1)}(0) = 1$$
$$f^{(2)} = -\sin x \qquad f^{(2)}(0) = 0$$
$$f^{(3)} = -\cos x \qquad f^{(3)}(0) = -1$$
$$f^{(4)} = \sin x \qquad f^{(4)}(0) = 0$$
$$f^{(5)} = \cos x \qquad f^{(5)}(0) = 1$$
$$f^{(6)} = -\sin x \qquad f^{(6)}(0) = 0$$

$$f^{(7)} = -\cos x \qquad f^{(7)}(0) = -1$$
$$f^{(8)} = \sin x \qquad f^{(8)}(0) = 0$$
$$f^{(9)} = \cos x \qquad f^{(9)}(0) = 1$$
$$f^{(10)} = -\sin x \qquad f^{(10)}(0) = 0$$
$$\vdots \qquad\qquad \vdots$$

Taylor series expansion for different orders are

$$f(x) = x + O(\Delta x^2),$$
$$f(x) = x - \frac{x^3}{3!} + \frac{x^5}{5!} - \frac{x^7}{7!} + O(\Delta x^8),$$
$$f(x) = x - \frac{x^3}{3!} + \frac{x^5}{5!} - \frac{x^7}{7!} + \frac{x^9}{9!} + O(\Delta x^{10}).$$

Figure 1.2 shows the graph of Taylor series and the graph of the sine function.

While the approximated functions wander off to infinity, retaining sufficient amount of terms will ensure the approximated function

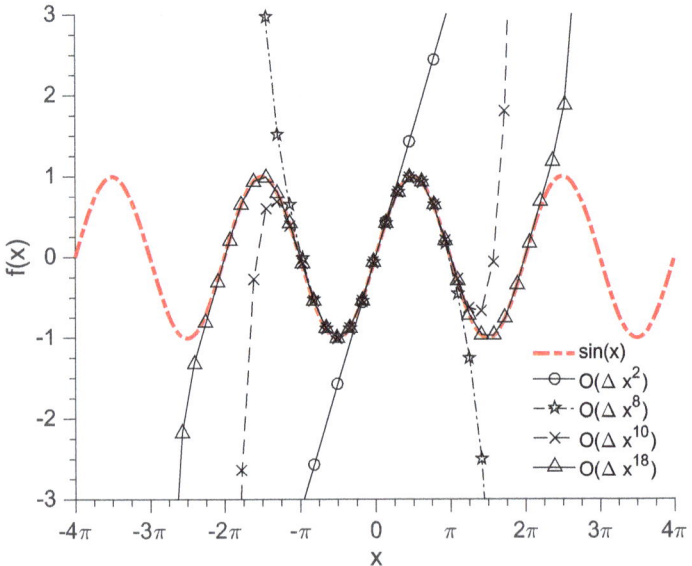

Figure 1.2:   Plot of function $\sin x$ and its successive Taylor series expansions.

to closely represent the sine function around the origin ($x = 0$). Taylor series expansion up to the 17th degree can recover the sine function accurately from -2π to 2π. If we were to retain more terms of the Taylor series expansion, we expect to recover the sine function accurately from -50π to 50π or -100π to 100π.

One caveat must be made: The expansion of all functions does not behave as well as the sine function. In fact, increasing the degree of expansion of function $\frac{1}{1-x}$ will not help its divergence for $|x| > 1$.

Example 4: Expand $e^{0.1x}$ and show the error associated with approximating the function.

Solution:

$$y = e^x = 1 + 0.1x + 0.01x^2/2 + 0.001x^3/3! + 0.0001x^4/4! + \cdots .$$

$$(1.17)$$

- Zero order: $y_0 = e^{0.1x} = 1$, truncation order $O(x)$,
- First order: $y_1 = e^{0.1x} = 1 + 0.1x$, truncation order $O(x^2)$,
- Second order: $y_2 = e^{0.1x} = 1 + 0.1x + 0.01x^2/2$, truncation order $O(x^3)$,
- and so on.

Figure 1.3 shows the plot of the function $e^{0.1x}$ and approximation of the function for different orders of accuracy at $x = 0$. Obviously, more terms are needed as the value of x distances from the point of expansion ($x = 0$).

1.5.2 Multi-variable functions

In most engineering and science applications, the dependent variable is a function of many variables. For instance, in thermodynamics, gas enthalpy and density are functions of temperature and pressure; in a turbulent flow in a pipe, the pressure drop is a function of Reynolds number and surface roughness of the pipe; and in convective heat transfer, Nusselt number is a function of Reynolds and Prandtl numbers. In general, assume that a function, $f(x, y)$, depends on x and y. Also, assume that the value of function is known at $x = x_0$ and

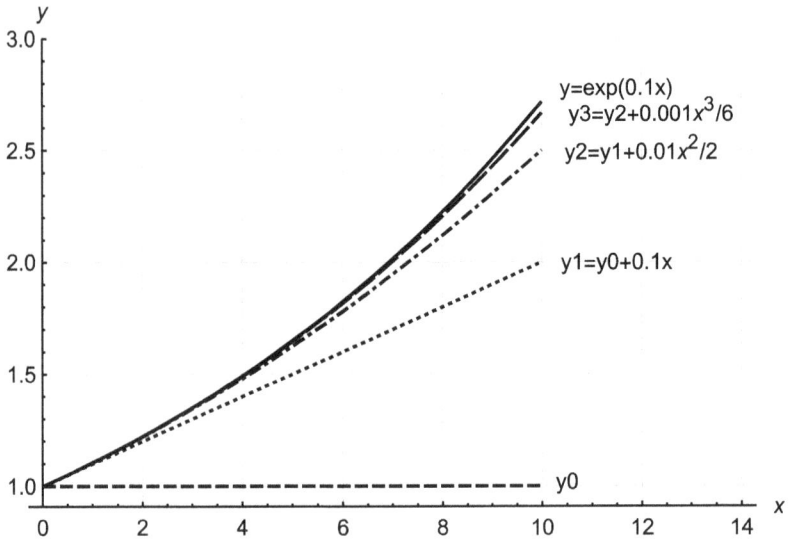

Figure 1.3:   Plot of function $e^{0.1x}$ and its successive Taylor series expansions.

$y = y_0$. Therefore, Taylor series expansion around this point is

$$f(x, y) = f(x_0, y_0) + \left.\frac{\partial f}{\partial x}\right|_{x_0, y_0} (x - x_0) + \left.\frac{\partial f}{\partial y}\right|_{x_0, y_0} (y - y_0)$$

$$+ \left.\frac{\partial^2 f}{\partial x \partial y}\right|_{x_0, y_0} (x - x_0)(y - y_0) + \frac{1}{2}\left.\frac{\partial^2 f}{\partial x^2}\right|_{x_0, y_0} (x - x_0)^2$$

$$+ \frac{1}{2}\left.\frac{\partial^2 f}{\partial y^2}\right|_{x_0, y_0} (y - y_0)^2 + \cdots . \tag{1.18}$$

A continuous function,

$$f(x, y) = \cos x \cos y, \tag{1.19}$$

has a shape as depicted in Fig. 1.4 and can be approximated using Taylor series about $(0, 0)$ as

$$f(x, y) = 1 - \frac{x^2}{2} - \frac{y^2}{2} + O(x^3, y^3). \tag{1.20}$$

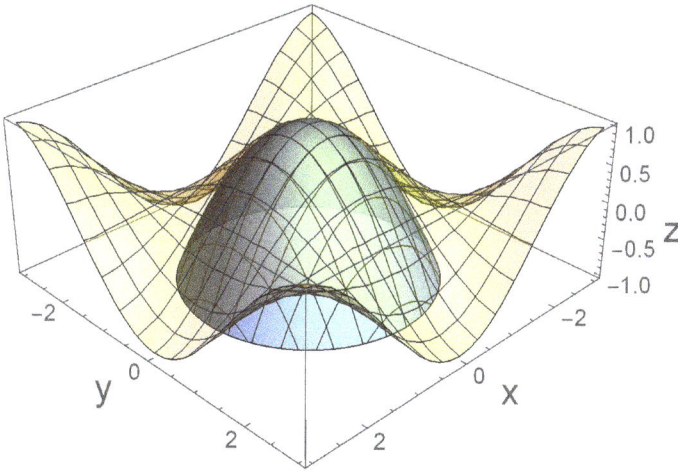

Figure 1.4: Plot of function $\cos x \cos y$ (yellow) and its second-order Taylor series expansion (blue) surfaces.

1.6 Problems

Problem 1: A black body at temperature T (K) emits electromagnetic radiation with density function (called Planck's law) as a function of wave length, λ, which is given by

$$f(\lambda) = \frac{8\pi hc}{\lambda^5 (e^{hc/\lambda kT} - 1)}, \tag{1.21}$$

where h, c, and k are Planck's constant, speed of light, and Boltzmann's constant, respectively; hc/k is about $0.014 \, \mathrm{m\,K}$. Using Taylor series, expand the exponential term up to the seventh term and determine the wavelength of the peak emission. Hint: Maximize $f(\lambda)$ by minimizing the quantity $\lambda^5 (e^{hc/\lambda kT} - 1)$. Compare your results to those of Wien's law ($\lambda_{\max} = \frac{0.002898}{T}$). Also, show that $f(\lambda)$ can be approximated as $\frac{8\pi kT}{\lambda^4}$.

Problem 2: Express $e^{I\,x}$, where $I^2 = -1$, in terms of $\sin x$ and $\cos x$.

Problem 3: Nonlinear Poisson–Boltzmann equation for a one-dimensional problem can be written as

$$\frac{d^2\phi}{dy^2} = -A \, e^{-z\phi}. \tag{1.22}$$

Table 1.1:   Taylor series of function $\ln(1+x)$ about $x = 0$.

x	$\ln(1+x)$	Using expansion up to the 5th term	Using expansion up to the 7th term
$\dfrac{1}{2}$			
2			

The equation has importance in understanding ions distribution in microchannels. A and z are constants. For small $z\phi$, simplify the above equation by linearizing it to the first order and solve the linear second-order ordinary equation.

Problem 4: Boussinesq approximation of buoyancy force for density as a function of temperature is $g\beta\Delta T$. Prove that. Note that β is the thermal expansion coefficient, where $\beta = \left.\dfrac{d\rho}{dT}\right|_{T=T_0}$. Expand ρ at T_0, $\Delta T = (T - T_0)$.

Problem 5: Using Taylor series approximation, integrate the following:

$$\int_0^1 \frac{e^x}{\sqrt{x}}dx. \tag{1.23}$$

Problem 6: Perform Taylor series expansion to approximate function $\ln(1+x)$ around $x = 0$ up to the seventh term. Fill in Table 1.1 with the exact value of $\ln(1+x)$ for $x = \frac{1}{2}$, 2 and its approximation using the expansion up to the 5th and 7th terms.

   Conclude about the convergence/divergence of the series for $x = \frac{1}{2}$ and $x = 2$ and the number of terms considered in Taylor series expansion.

Problem 7: Perform Taylor series expansion of function $\frac{1}{1+\sin^2 x}$ around $x = 0$ up to the third term. (Hint: Since $\sin 0 = 0$, let $y = \sin^2 x$, expand $\frac{1}{1+y}$ around $y = 0$, then expand $y = \sin^2 x$ around $x = 0$.)

Extra Reading Material: First-year calculus.

Chapter 2

Linear Algebra

2.1 Linear Algebra

Linear algebra is one of the main ingredients of numerical analysis. In general, solving differential equations (ordinary, ODEs, and partial, PDEs) numerically ends up with solving a system of linear algebraic equations. In fact, the bottleneck of solving differential equations (DEs) numerically is the computer time consumed in solving the algebraic equations, which are the outcome of approximations of the DEs. In the following sections, we touch upon the basics of linear algebra techniques that will be used later on.

2.2 Vectors and Matrices

2.2.1 Definitions and properties

A vector is a one-dimensional (row or column) array, usually represented as

$$\mathbf{A} = \sum_{i=1}^{n} a_i \, \mathbf{e}_i,$$

where $\mathbf{e}_i$ is a unit vector (that belongs to a basis in which this vector is represented), coefficient a_i is the magnitude (or more accurately, the algebraic measure or cosine coefficient) of the vector projection in the e_i direction, and n is the size (or dimension) of $\mathbf{A}$. For example,

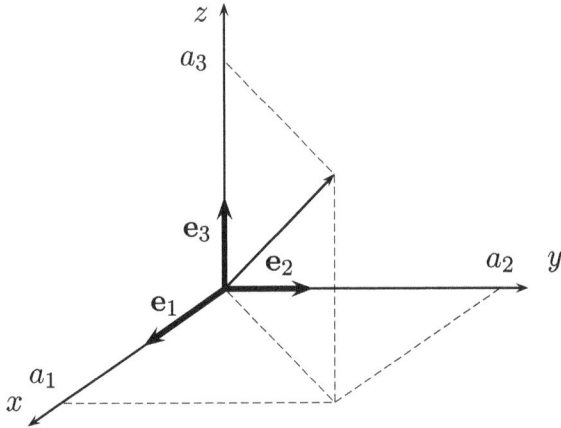

Figure 2.1:   Sketch of a three-dimensional vector.

for a vector in three-dimensional space, we write $\mathbf{A} = a_1\,\mathbf{e}_1 + a_2\,\mathbf{e}_2 + a_3\,\mathbf{e}_3$, see Fig. 2.1. $\mathbf{e}_1$, $\mathbf{e}_2$, and $\mathbf{e}_3$ are unit vectors in x, y, and z directions, respectively. Also, the row vector can be written as

$$\mathbf{A} = [a_1,\ a_2,\ a_3,\ a_4,\ \ldots,\ a_n].$$

It is more convenient to represent a column vector as the transpose of a row vector, as

$$\mathbf{B} = [b_1,\ b_2,\ b_3,\ b_4,\ \ldots,\ b_n]^T,$$

mainly to save space instead of writing the vector vertically.

The sum of two vectors A and B, of the same size n, is the vector C of size n and denoted by $A + B$, given by

$$c_i = a_i + b_i,$$

where a_i, b_i, and c_i are elements of vectors A, B, and C, respectively. The scalar multiplication of a scalar λ and a vector A, of size n, is the vector B of size n and denoted by λA, given by

$$b_i = \lambda\,a_i \quad i = 1,\ldots,n.$$

The dot (or inner) product of vectors $\mathbf{A}$ and $\mathbf{B}$, denoted $\mathbf{A} \cdot \mathbf{B}$, is the scalar ϕ given by

$$\phi = \sum_{i=1}^{n} a_i\, b_i. \tag{2.1}$$

Usually, the length (or magnitude or norm) of a vector is represented as $\|\mathbf{A}\|$. The Euclidean length of a column vector is given by $\sqrt{\mathbf{A}^T \cdot \mathbf{A}}$. For a three-dimensional vector $\mathbf{A}$, it reads $\sqrt{a_1^2 + a_2^2 + a_3^2}$.

Matrix: A matrix is a combination of vectors, i.e., a column vector of raw vectors (or vice versa) of the same size. It can be a rectangular or a square matrix. A matrix having m number of rows and n number of columns is called $m \times n$ matrix, in which case (m, n) is called the size of the matrix. For example, a system of linear algebraic equations can be represented as

$$
\begin{aligned}
a_{11}x_1 \quad & a_{12}x_2 \quad \cdots + a_{1n}x_n = b_1, \\
a_{21}x_1 \quad & a_{22}x_2 \quad \cdots + a_{2n}x_n = b_2, \\
\vdots \qquad & \vdots \qquad \ddots + \quad \vdots \quad = \vdots \\
a_{m1}x_1 \quad & a_{m2}x_2 \quad \cdots + a_{mn}x_n = b_m.
\end{aligned}
$$

Each of the above equations can be written as a dot product of vectors:

$$
\begin{aligned}
[a_{11} \quad a_{12} \quad a_{13} \quad \cdots][x_1 \quad x_2 \quad x_3 \quad \cdots]^T &= b_1, \\
[a_{21} \quad a_{22} \quad a_{23} \quad \cdots][x_1 \quad x_2 \quad x_3 \quad \cdots]^T &= b_2, \\
\vdots
\end{aligned}
\tag{2.2}
$$

and so on.

The shorthand form of the above equations is $AX = B$, where

$$
\mathbf{A} = \begin{bmatrix} a_{11} & a_{12} & \cdots & a_{1n} \\ a_{21} & a_{22} & \cdots & a_{2n} \\ \vdots & \vdots & \ddots & \vdots \\ a_{m1} & a_{m2} & \cdots & a_{mn} \end{bmatrix}, \quad \mathbf{B} = \begin{bmatrix} b_1 \\ b_2 \\ \vdots \\ b_n \end{bmatrix}, \quad \mathbf{X} = \begin{bmatrix} x_1 \\ x_2 \\ \vdots \\ x_n \end{bmatrix},
$$

where $\mathbf{A}$ is a square matrix, that is $m = n$, and $\mathbf{B}$ and $\mathbf{X}$ are column vectors of the same size n. The sum of the two matrices $\mathbf{A}$ and $\mathbf{B}$ of the same size (m, n), denoted $\mathbf{A} + \mathbf{B}$, is the $m \times n$ matrix $\mathbf{C}$ given by

$$
c_{ij} = a_{ij} + b_{ij}, \qquad 1 \leqslant i \leqslant m, \quad 1 \leqslant j \leqslant n,
$$

where a_{ij}, b_{ij}, and c_{ij} are elements of matrices $\mathbf{A}$, $\mathbf{B}$, and $\mathbf{C}$, respectively. So, matrix addition is commutative, i.e., $\mathbf{A} + \mathbf{B} = \mathbf{B} + \mathbf{A}$, and associative, i.e., $(\mathbf{A} + \mathbf{B}) + \mathbf{C} = \mathbf{A} + (\mathbf{B} + \mathbf{C})$.

The product of the two matrices $\mathbf{A}$, of size (m, n), and $\mathbf{B}$, of size (n, p), denoted $\mathbf{A} \cdot \mathbf{B}$ or simply $\mathbf{A}\,\mathbf{B}$, is the $m \times p$ matrix $\mathbf{C}$ given by

$$c_{ij} = \sum_{k=1}^{n} a_{ik}\, b_{kj}, \qquad 1 \leqslant i \leqslant m, \quad 1 \leqslant j \leqslant p.$$

In general, matrix multiplication is not commutative; that is, usually, $\mathbf{A} \cdot \mathbf{B} \neq \mathbf{B} \cdot \mathbf{A}$. However, it is associative, i.e., $(\mathbf{A} \cdot \mathbf{B}) \cdot \mathbf{C} = \mathbf{A} \cdot (\mathbf{B} \cdot \mathbf{C})$.

Diagonal elements refer to elements on the diagonal of a square matrix. The row and column indices of each of these elements are equal.

The identity matrix, commonly denoted $\mathbf{I}$, is a matrix whose elements are all zero except the diagonal elements that equal to 1:

$$\begin{bmatrix} 1 & 0 & 0 & \cdots & 0 \\ 0 & 1 & 0 & \cdots & 0 \\ \multicolumn{5}{c}{\dotfill} \\ 0 & 0 & 0 & \cdots & 1 \end{bmatrix}.$$

The upper triangular matrix, usually denoted $\mathbf{U}$, is a square matrix whose elements below the diagonal are all zero, i.e.,

$$\mathbf{U}_{ij} \quad \text{for all } i > j, \ j = 1, \ldots, n.$$

The lower triangular matrix, usually denoted $\mathbf{L}$, is a square matrix whose elements above the diagonal are all zero, i.e.,

$$\mathbf{U}_{ij} \quad \text{for all } i < j, \ j = 1, \ldots, n.$$

The sub-matrix $\mathbf{A}\ (n-1 \times m-1; p, q)$ of an $n \times m$ matrix $\mathbf{A}$ is the matrix obtained by removing row p and column q from matrix $\mathbf{A}$. It is usually denoted by $\overline{\mathbf{A}}_{\{p,q\}}$.

The trace of an $n \times n$ (and so square) matrix $\mathbf{A}$, denoted $\mathrm{tr}(\mathbf{A})$, is the scalar given by the sum of all diagonal elements of matrix $\mathbf{A}$, i.e.,

$$\mathrm{tr}(\mathbf{A}) = \sum_{k=1}^{n} a_{ii}.$$

The determinant of an $n \times n$ matrix $\mathbf{A}$, denoted $\det(\mathbf{A})$ or $|\mathbf{A}|$, is the scalar given by

$$\det(\mathbf{A}) = \sum_{k=1}^{n}(-1)^{k-1}a_{1,k}\det(\overline{\mathbf{A}}_{\{1,k\}}).$$

By convention, the determinant of a 1×1 matrix is its unique element itself.

The determinant of upper and lower triangular matrices equals to the product of their diagonal elements. A matrix whose determinant is zero is called singular.

The inverse of a non-singular square matrix $\mathbf{A}$, denoted $\mathbf{A}^{-1}$, is such that $\mathbf{A} \cdot \mathbf{A}^{-1} = \mathbf{A}^{-1} \cdot \mathbf{A} = \mathbf{I}$.

The transpose of a matrix $\mathbf{A}$ is $\mathbf{A}^T$, where $\mathbf{A}^T$ can be obtained by switching its rows' elements with its columns' elements.

Remark: Any matrix $\mathbf{B}$ can be written as a sum of a symmetric and a skew-symmetric matrix:

$$\mathbf{B} = \tfrac{1}{2}\left(\mathbf{B} + \mathbf{B}^T\right) + \tfrac{1}{2}\left(\mathbf{B} - \mathbf{B}^T\right).$$

The **tri-diagonal matrix** is a banded square matrix, where all elements are zero except the diagonal and two adjacent elements to diagonal elements:

$$\begin{bmatrix} b_1 & c_1 & 0 & 0 & \cdots & 0 & 0 \\ a_2 & b_2 & c_2 & 0 & \cdots & 0 & 0 \\ 0 & a_3 & b_3 & c3 & 0 & \cdots & 0 \\ 0 & 0 & a_4 & b_4 & c_4 & \cdots & 0 \\ \cdots & \cdots & \cdots & \cdots & \cdots & \cdots & \cdots \\ 0 & 0 & 0 & \cdots & 0 & a_n & b_n \end{bmatrix}.$$

In general, to save computer memory, the tri-diagonal matrix is stored as three vectors:

$$[a] = [0 \; a_2 \; a_3 \; a_4 \; \cdots \; a_{n-1}]^T,$$
$$[b] = [b_1 \; b-2 \; b_3 \; b_4 \; \cdots \; b_n]^T,$$
$$[c] = [c_1 \; c_2 \; c_3 \; c_4 \; \cdots \; c_{n-1}, 0]^T.$$

Euclidean norm: The resulting double dot product of a square matrix $\mathbf{A}$ is called L^2-norm or Euclidean norm:

$$\|\mathbf{A}\| = \sqrt{\operatorname{tr}(\mathbf{A}^T\mathbf{A})} = \sqrt{\sum_{i=1}^{n} a_i^2}. \tag{2.3}$$

The Euclidean norm of the residual vector is used to monitor convergence of the iterative solution, see Section 2.6.

On the other hand, L^1-norm is $\sum_{i=1}^{n} |a_{ii}|$.

Symmetric matrix: $\mathbf{A}^T = \mathbf{A}$.

Skew-symmetric matrix: $\mathbf{A}^T = -\mathbf{A}$.

Note: $(\mathbf{A}\,\mathbf{B})^T = \mathbf{B}^T\,\mathbf{A}^T$ and $(\mathbf{A}\,\mathbf{B})^{-1} = \mathbf{B}^{-1}\,\mathbf{A}^{-1}$.

Eigenvalues and eigenvectors: Let $\mathbf{A}$ be a n-square matrix. If there exists a nonzero scalar λ and a vector V such that

$$\mathbf{A}\,V = \lambda\,V, \tag{2.4}$$

then λ is called the eigenvalue of matrix $\mathbf{A}$ associated with eigenvector $\mathbf{V}$. A matrix $\mathbf{A}$ has at most n different eigenvalues. To obtain all eigenvalues of $\mathbf{A}$, we have to solve the following equation:

$$\det(\mathbf{A} - \lambda\mathbf{I}) = 0.$$

According to (2.4), multiplying matrix $\mathbf{A}$ by a vector $\mathbf{X}$ is equivalent to stretching (or squeezing) vector $\mathbf{X}$ by a factor $|\lambda_1|$ in the direction of vector $\mathbf{V}_1$, by a factor $|\lambda_2|$ in the direction of vector $\mathbf{V}_2$, etc., where λ_1, λ_2, ... are the successive eigenvalues of $\mathbf{A}$.

Spectral radius: The spectral radius of a matrix $\mathbf{A}$ is given by the value of its largest absolute eigenvalue. The spectral radius specifies the maximum stretch factor applied to a given vector $\mathbf{X}$ when the latter is multiplied by matrix $\mathbf{A}$.

Singular values: The singular values of a matrix $\mathbf{A}$ are the square roots of the eigenvalues of matrix $\mathbf{A}^T\mathbf{A}$.

Condition number: The condition number of a matrix **A** is given by

$$\frac{\max_i |\sigma_i|}{\min_i |\sigma_i|},$$

where σ_i are the singular values of matrix **A**.

The condition number of a matrix **A** is a measure of "how far" from the identity matrix is matrix **A**. This number is necessarily greater than or equal to one. The larger the condition number, the farther away the matrix from its identity matrix. The point is that multiplication of any vector **X** by the identity matrix is trivial and so not prone to any round-off errors, while the multiplication of this vector by any other matrix is usually subjected to such errors. Consequently, the larger the condition number, the more prone the matrix inversion (or equivalently solving problems involving this very same matrix) is to such errors and the lower the accuracy of the result.

Diagonal dominance: If the absolute value of the diagonal element is greater than or equal to the sum of absolute values of all other elements on a row, the matrix is called **diagonally dominant matrix**, provided that the condition applies to every row:

$$\text{for all } i, \quad |a_{ii}| \geqslant \sum_{\substack{j=1 \\ j \neq i}}^{n} |a_{ij}|. \tag{2.5}$$

When the inequality in (2.5) is strict, the matrix is called strictly diagonally dominant. If the matrix is not diagonally dominant, then it is called **ill-conditioned matrix**.

Positive and negative definite: A matrix **A** is positive definite if $X^T \mathbf{A} X > 0$, where X is any nonzero vector. Conversely, if $X^T \mathbf{A} X < 0$, for all nonzero vector X, then **A** is said to be negative definite. The negativeness of a matrix **A** is useful to ascertain the stability of systems of differential equations of the type $\frac{d\mathbf{X}}{dt} = \mathbf{A}\mathbf{X}$.

Positive and negative semi-definite: A matrix **A** is positive semi-definite if $X^T \mathbf{A} X \geqslant 0$, where X is any nonzero vector. Conversely, if $X^T AX \leqslant 0$, for all nonzero vector X, then $\mathbf{A}$ is called negative semi-definite.

Orthogonal vector: If the product of two vectors, say $\mathbf{X}$ and $\mathbf{Z}$, is zero, i.e., $\mathbf{X}^T \mathbf{Z} = 0$, then these vectors are said to be orthogonal.

2.2.2 Examples

The following examples illustrate the above concepts.

(a) Matrix $\mathbf{A}$ given by

$$\mathbf{A} = \begin{bmatrix} 1 & 2 & -6 \\ 2 & 4 & 9 \\ -6 & 9 & 3 \end{bmatrix}$$

is symmetric, while matrix $\mathbf{B}$ given by

$$\mathbf{B} = \begin{bmatrix} 0 & 2 & -6 \\ -2 & 0 & 9 \\ 6 & -9 & 0 \end{bmatrix}$$

is skew-symmetric.

(b) Consider matrix $\mathbf{D}$ which arises in one-dimensional diffusion problems. This matrix usually reads

$$\mathbf{D} = \begin{bmatrix} 2 & -1 & 0 \\ -1 & 2 & -1 \\ 0 & -1 & 2 \end{bmatrix}.$$

Then, consider a vector $\mathbf{x} = (x_1, x_2, x_3)^T$ of real components not all equal to zero. The scalar $\mathbf{x}^T \mathbf{D} \mathbf{x}$ is given by

$$2[x_1^2 + x_2^2 + x_3^2 - x_1 x_2 - x_2 x_3]$$
$$= 2\left[(x_1 - \tfrac{1}{2}x_2)^2 + (x_3 - \tfrac{1}{2}x_2)^2 + \tfrac{1}{2}x_2^2\right],$$

which is always positive. Therefore, matrix $\mathbf{D}$ is positive definite.

(c) The vectors $\mathbf{V}_1 = (\frac{\sqrt{3}}{2}, \frac{1}{2}, 1)^T$ and $\mathbf{V}_2 = (-\frac{1}{2}, \frac{\sqrt{3}}{2}, 0)^T$ are orthogonal since $\mathbf{V}_1^T \cdot \mathbf{V}_2 = -\frac{\sqrt{3}}{2}\frac{1}{2} + \frac{1}{2}\frac{\sqrt{3}}{2} + 0 = 0$.

(d) Let us compute the eigenvalues and eigenvectors of matrix $\mathbf{C}$ given by

$$\mathbf{C} = \begin{bmatrix} -\frac{1}{2} & \frac{\sqrt{3}}{2} & 0 \\ \frac{\sqrt{3}}{2} & \frac{1}{2} & 0 \\ 0 & 0 & 2 \end{bmatrix}.$$

Subtract $\lambda\mathbf{I}$ from the above matrix $\mathbf{C}$:

$$\mathbf{C} - \lambda\mathbf{I} = \begin{bmatrix} -\frac{1}{2} - \lambda & \frac{\sqrt{3}}{2} & 0 \\ \frac{\sqrt{3}}{2} & \frac{1}{2} - \lambda & 0 \\ 0 & 0 & 2 - \lambda \end{bmatrix}.$$

An eigenvalue λ of matrix $\mathbf{C}$ satisfies the equation

$$\det(\mathbf{C} - \lambda\mathbf{I}) = (2 - \lambda)\left(\lambda^2 - 1\right) = 0.$$

The roots of this equation, -1, 1, and 2, are the eigenvalues of matrix $\mathbf{C}$. To get an eigenvector associated with the eigenvalue $\lambda = 1$, for instance, we have to solve the following system of equations:

$$\mathbf{CV}_\lambda = \lambda\mathbf{V}_\lambda,$$

that is, if $\mathbf{V}_\lambda = (V_x, V_y, V_z)^T$,

$$\begin{cases} \frac{1}{2}\left(-V_x(4\lambda + 1) + V_y\left(-\sqrt{3}\right)\right) = 0 & \frac{1}{2}\left(-V_x(4\lambda + 1) + V_y\left(-\sqrt{3}\right)\right) = 0 \\ \frac{1}{2}\left(-4\lambda V_y + V_y + V_x\left(-\sqrt{3}\right)\right) = 0 & \Leftrightarrow \frac{1}{2}\left(-4\lambda V_y + V_y + V_x\left(-\sqrt{3}\right)\right) = 0 \\ -2V_z(\lambda - 1) = 0 & 0 = 0. \end{cases}$$

If we substitute λ by 1 in this system, we find that

$$\begin{cases} V_x & - & 0, \\ V_y & = & 0, \\ V_z & = & \text{any nonzero real number.} \end{cases}$$

So, an eigenvector associated with the eigenvalue $\lambda = 1$ can be $(0, 0, 1)^T$.

The singular values of matrix $\mathbf{C}$ are the eigenvalues of $\mathbf{C}^T \cdot \mathbf{C}$. It can be checked easily that they are given by 1, 1, and 4.

Thus, the spectral radius of matrix $\mathbf{C}$ is $\max(|-1|, |1|, |2|) = 2$ and its conditioning is $\dfrac{\max(|1|, |1|, |4|)}{\min(|1|, |1|, |4|)} = 4.$

2.3   System of Linear Algebraic Equations

Approximating differential equations usually leads to a system of algebraic equations. A linear system can be written as

$$\mathbf{A}\,\mathbf{x} = \mathbf{b}, \tag{2.6}$$

where $\mathbf{A}$ is $n \times n$ non-singular square matrix and $\mathbf{x}$ and $\mathbf{b}$ are vectors. Vector $\mathbf{x}$ is unknown and vector $\mathbf{b}$ includes external driving force term. The solution is $\mathbf{x} = \mathbf{A}^{-1}\mathbf{b}$.

In general, there are two classes of methods to solve a system of linear algebraic equations:

- Direct methods,
- Iterative methods.

Remark: For systems involving diagonally dominant matrices, both mentioned classes lead to a good approximation of the solution (and a convergent solution for the latter too). For ill-conditioned matrices, there is no guarantee of accuracy and/or convergence and the method of solution should be carefully selected. Hence, round-off errors will have effects on the convergence of the solution. For iterative methods, the initial guesses have an effect on the number of iterations.

An example to illustrate the effect of round-off error on an ill-conditioned matrix is the following:

$$x + 2y = 6, \quad x + 2.001y = 7.$$

The solution is $x = -1994$ and $y = 1000$. However, changing (slightly) the coefficient of y from 2.001 to 2.0012 gives $x = 1660.66$ and $y = 833.33$. The solution is exceedingly sensitive to the problem's input.

2.4   Direct Methods

To solve a system of linear equations $\mathbf{Ax} = \mathbf{B}$, where $\mathbf{A}$ is an $n \times n$ matrix and $\mathbf{b}$ is a known vector of size n, it is possible in theory to use Cramer's rule given by the relationship,

$$x_i = \frac{\det_i(\mathbf{A})}{\det(\mathbf{A})}, \quad i = 1, \ldots, n, \tag{2.7}$$

where $\det_i(\mathbf{A})$ is the determinant of matrix $\mathbf{A}$ when its ith column is replaced by vector $\mathbf{b}$. Numerically, this method is absolutely not usable: It leads to a large number of mathematical operations (multiplications and divisions). It can be shown that this number is about $n(n + 1)!$. This strategy should therefore be avoided except for systems of very small size (10 or so).

Exact solution can be achieved with other direct methods, assuming that the round-off errors are negligible. However, they are not efficient in solving systems of large number of equations, in order of 1000 and more. For systems of smaller number of equations, they can be used safely. Let us present **Gauss elimination (GE) method**, which is the basis of most direct methods.

2.4.1 Gauss elimination method

GE method aims at transforming $\mathbf{A}$ into an upper diagonal matrix by systematic process of elimination. Through the elimination process, vector $\mathbf{b}$ is modified too. The solution for the vector $\mathbf{x}$ can be obtained simply by back substitutions. Since, for a given step, the elimination process of each row does not affect other rows, it is possible to pipe each process in a different computer processor to accelerate the solution. The algorithm for solving $\mathbf{Ax} = \mathbf{B}$ is as follows:

- Read the coefficients of the $n \times n$ matrix, a_{ij}, and the n vector, b_i, where n is the number of rows and number of columns, a_{ij}, $i, j = 1, \ldots, n$ are elements of matrix $\mathbf{A}$, and b_i, $i = 1, \ldots, n$ are elements of vector $\mathbf{B}$. i and j are indices of row and column, respectively.
- For column k $(k = 1, 2, 3, \ldots, n - 1)$ and row i $(i = k + 1, k + 2, \ldots, n)$, this process will update the right-hand side vector and create an upper triangular matrix at the same time:

$$\begin{cases} b_i = b_i - \left(\dfrac{a_{ik}}{a_{kk}} \right) b_k \\[3mm] a_{ij} = a_{ij} - \left(\dfrac{a_{ik}}{a_{kk}} \right) a_{kj} \quad (j = 1, \ldots, n) \end{cases} \qquad (i = k + 1, \ldots, n).$$

$$(2.8)$$

- Back substitutions process: To solve for vector **x**,

$$x_n = \frac{b_n}{a_{nn}}, \tag{2.9}$$

$$x_i = \frac{b_i - \sum_{j=i+1}^{n} a_{ij} x_j}{a_{ij}} \quad i = n-1, n-2, \ldots, 1. \tag{2.10}$$

Note: The number of mathematical operations (multiplications and divisions) for GE is about $n^3/3 - n/3$, i.e., $O(n^3/3)$ for large n. For 1000 equations $(n = 1000)$, which is very common in engineering applications, the number of operation is more than 3×10^8. Therefore, GE is highly inefficient in solving systems of large number of equations (typically over 1000).

Example: Let us consider the problem $\mathbf{AX} = \mathbf{B}$, where $\mathbf{B} = (-1, -1, 1)^T$ and $\mathbf{A}$ is one of the following matrices:

$$\mathbf{A_1} = \begin{bmatrix} 6 & 5 & 4 \\ 12 & 13 & 10 \\ 28 & 27 & 21 \end{bmatrix}, \quad \mathbf{A_2} = \begin{bmatrix} 1 & 1 & 3 \\ 2 & 2 & 2 \\ 3 & 6 & 4 \end{bmatrix},$$

$$\text{or} \quad \mathbf{A_3} = \begin{bmatrix} 1 & \left\{ \begin{matrix} 1+ \\ 2\times 10^{-7} \end{matrix} \right\} & 3 \\ 2 & 2 & 2 \\ 3 & 6 & 4 \end{bmatrix}.$$

Let us apply GE using matrix $\mathbf{A_1}$ and the right-hand side vector $\mathbf{B}$:

$$\begin{bmatrix} 6 & 5 & 4 & | & -1 \\ 12 & 13 & 10 & | & -1 \\ 28 & 27 & 21 & | & 1 \end{bmatrix} \Rightarrow \begin{bmatrix} 6 & 5 & 4 & | & -1 \\ 0 & 3 & 2 & | & 1 \\ 0 & 3.666 & 2.333 & | & 5.666 \end{bmatrix}$$

$$\Rightarrow \begin{bmatrix} 6 & 5 & 4 & | & -1 \\ 0 & 3 & 2 & | & 1 \\ 0 & 0 & -0.111 & | & 4.444 \end{bmatrix}.$$

A back substitution yields the solution: $\mathbf{X} = (4, 27, -40)^T$. The residual $\mathbf{B} - \mathbf{AX} = (0, 0, 0)^T$ is excellent which means that the solution found is accurate.

Let us repeat the same procedure with matrix $\mathbf{A}_2$:

$$
\begin{bmatrix}
1 & 1 & 3 & -1 \\
2 & 2 & 2 & -1 \\
3 & 6 & 4 & 1
\end{bmatrix}
\Rightarrow
\begin{bmatrix}
1 & 1 & 3 & -1 \\
0 & 0 & -4 & 1 \\
0 & 3 & -5 & 4
\end{bmatrix}
$$

$$
\Rightarrow
\begin{bmatrix}
1 & 1 & 3 & -1 \\
0 & 0 & -4 & 1 \\
\text{NAN} & \text{NAN} & \text{Inf} & -\text{Inf}
\end{bmatrix}.
$$

Not-a-number (NAN) and **infinity** symbols are produced, which means something has gone wrong with the calculation. The problem arises when the pivot becomes zero (step 2, diagonal element of row 2, shown in red). So, GE fails if ever the pivot is zero. This prevents getting any solution. Therefore, other methods have to be used. The problem can even become worse if we consider a slightly different matrix, $\mathbf{A}_3$, in which case the steps proceed as follows:

$$
\begin{bmatrix}
1 & \left\{\begin{smallmatrix}1+\\2\times10^{-7}\end{smallmatrix}\right\} & 3 & -1 \\
2 & 2 & 2 & -1 \\
3 & 6 & 4 & 1
\end{bmatrix}
\Rightarrow
\begin{bmatrix}
1 & \left\{\begin{smallmatrix}1+\\2\times10^{-7}\end{smallmatrix}\right\} & 3 & -1 \\
0 & -4.768\times10^{-7} & -4 & 1 \\
0 & 2.999 & -5 & 4
\end{bmatrix}
$$

$$
\Rightarrow
\begin{bmatrix}
1 & \left\{\begin{smallmatrix}1+\\2\times10^{-7}\end{smallmatrix}\right\} & 3 & -1 \\
0 & -4.768\times10^{-7} & -4 & 1 \\
0 & 0 & -25165824 & 6291458.5
\end{bmatrix}.
$$

No error is displayed during the calculation and a back substitution yields the solution: $\mathbf{X} = (-1,\ 0.75,\ -0.25)^T$. However, the residual $\mathbf{B} - \mathbf{AX} = (0,\ 0,\ 0.5)^T$ is very bad. The most dangerous cases are those which do not generate any exception or errors and proceed silently and smoothly.

2.4.2 Thomas algorithm

In many engineering applications, we end up with a system of equations, which forms a tri-diagonal matrix. In such a case, the GE process needs to be carried out only on one element in each row below the main diagonal. For example,

$$
\mathbf{A} = \begin{bmatrix} a_{11} & a_{12} & 0 & 0 & 0 & 0 \\ a_{21} & a_{22} & a_{23} & 0 & \cdots & 0 \\ 0 & a_{32} & a_{33} & a_{34} & 0 & 0 \\ 0 & 0 & a_{43} & a_{44} & a_{45} & 0 \\ 0 & 0 & 0 & a_{54} & a_{55} & a_{56} \end{bmatrix}, \quad \mathbf{B} = \begin{bmatrix} b_1 \\ b_2 \\ b_3 \\ b_4 \\ b_5 \end{bmatrix}, \quad \mathbf{X} = \begin{bmatrix} x_1 \\ x_2 \\ x_3 \\ x_4 \\ x_5 \end{bmatrix}.
$$

Since most elements of tri-diagonal matrix are zeros, it saves a lot of computer memory by storing only nonzero diagonals elements, i.e., the elements of the main diagonal and the elements above and below the main diagonal, as

$$
\begin{bmatrix} 0 & a_{11} & a_{12} \\ a_{21} & a_{22} & a_{23} \\ a_{32} & a_{33} & a_{34} \\ a_{43} & a_{44} & a_{45} \\ a_{54} & a_{55} & a_{56} \\ .. & .. & .. \\ a_{n\,n-1} & a_{nn} & 0 \end{bmatrix} \quad \text{or} \quad \begin{bmatrix} 0 & d_1 & c_1 \\ a_2 & d_2 & c_2 \\ a_3 & d_3 & c_3 \\ a_4 & d_4 & c_4 \\ a_5 & d_5 & c_5 \\ \cdots & \cdots & \cdots \\ a_n & d_n & 0 \end{bmatrix}.
$$

The algorithm is as follows:

- Forward elimination:

$$
d_i = d_i - \left(\frac{a_i}{d_{i-1}}\right) c_{i-1}
$$
$$
b_i = b_i - \left(\frac{a_i}{d_{i-1}}\right) b_{i-1}
$$
$$(i = 2, 3, 4, \ldots, n). \tag{2.11}$$

- Backward substitution:

$$
x_n = \frac{b_n}{d_n}
$$
$$
x_i = \frac{b_i - c_i x_{i+1}}{d_i} \quad (i = n-1, n-2, \ldots, 2, 1). \tag{2.12}
$$

Note: The number of operations is about $5n$.

Example: Let us use Thomas algorithm to solve the problem $\mathbf{AX} = \mathbf{B}$, where

$$\mathbf{A} = \begin{bmatrix} -2 & 1 & 0 \\ 1 & -2 & 1 \\ 0 & 1 & -2 \end{bmatrix} \quad \text{and} \quad \mathbf{B} = \begin{bmatrix} 1 \\ 3 \\ -2 \end{bmatrix}.$$

- The forward elimination reads

$$\left[\begin{array}{ccc|c} -2 & 1 & 0 & 1 \\ 1 & -2 & 1 & 3 \\ 0 & 1 & -2 & -2 \end{array}\right] \Rightarrow \left[\begin{array}{ccc|c} -2 & 1 & 0 & 1 \\ 0 & -1.5 & 1 & 3.5 \\ 0 & 0 & -1.333 & 0.333 \end{array}\right].$$

- The back substitution gives

$$x_3 = -0.25, \quad x_2 = -2.5, \quad x_1 = -1.75.$$

The residual $\mathbf{B} - \mathbf{AX} = (0,\ 0,\ 0)^T$ is excellent.

2.4.3 Gauss–Jordan method

This method is a modification of GE method. The elimination process is the same as GE method carried out to convert matrix $\mathbf{A}$ into an upper triangular matrix. Also, the elimination process is carried out on the upper triangular matrix. The outcome of the elimination process is a diagonal matrix. The number of mathematical operations is higher than that of GE by n^2. Hence, the method is usually not recommended, but it may be used to calculate the inverse of a matrix.

2.4.4 LU factorization (decomposition)

In some cases, we have a linear system, such as an electrical network, and our objective is to study the response of the system by applying different voltages to it (forcing term). Also, in a water distribution network, the system of pipes network (matrix A) is fixed. However, we would like to test the system by applying different pressure gradients (vector b) on the flow rates (vector x) at each branch. The problem becomes one of solving a system of linear algebraic equations with different vectors $\mathbf{b}$. Since the matrix $\mathbf{A}$ is not changing, it is therefore beneficial to decompose $\mathbf{A}$ into a product of upper triangular matrix, $\mathbf{U}$, and lower triangular matrix, $\mathbf{L}$, by elimination processes as we did for GE ($\mathbf{A} = \mathbf{LU}$). Thus,

$$\mathbf{LUx} = \mathbf{b}. \tag{2.13}$$

Multiplying both sides of the above equation by inverse of $\mathbf{L}$ ($\mathbf{L}^{-1}$) yields

$$\mathbf{L}^{-1}\mathbf{LUx} = \mathbf{Ux} = \mathbf{L}^{-1}\mathbf{b}. \tag{2.14}$$

Let $\mathbf{L}^{-1}\mathbf{b} = \mathbf{b}'$, hence $\mathbf{b} = \mathbf{Lb}'$ and $\mathbf{Ux} = \mathbf{b}'$. The procedure is to first find $\mathbf{b}'$ by solving $\mathbf{Lb}' = \mathbf{b}$, then use $\mathbf{Ux} = \mathbf{b}'$ to determine $\mathbf{x}$ vector. It is worth noting here that calculating the inverse matrices of $\mathbf{L}$ and $\mathbf{U}$ is not needed since a forward (resp. back) substitution is used in practice to get $\mathbf{b}'$ (resp. $\mathbf{x}$).

One method of constructing $\mathbf{LU}$ decomposition is using regular GE method to construct $\mathbf{U}$ matrix. The $\mathbf{L}$ matrix elements below the diagonal are replaced by a_{ij}/a_{ii} for each i. The diagonal elements are set to unity.

Example: Let matrix $\mathbf{A} = \begin{bmatrix} 80 & -20 & -20 \\ -20 & 40 & -20 \\ -20 & -20 & 130 \end{bmatrix}$.

The $\mathbf{U}$ and $\mathbf{L}$ matrices are

$$\mathbf{U} = \begin{bmatrix} 80 & -20 & -20 \\ 0 & 40 & -20 \\ 0 & -20 & 130 \end{bmatrix} \Rightarrow \begin{bmatrix} 80 & -20 & -20 \\ 0 & 35 & -25 \\ 0 & 25 & -125 \end{bmatrix} \Rightarrow \begin{bmatrix} 80 & -20 & -20 \\ 0 & 35 & -25 \\ 0 & 0 & 750/7 \end{bmatrix},$$

$$\mathbf{L} = \begin{bmatrix} 1 & 0 & 0 \\ -20/80 & 1 & 0 \\ -20/80 & -25/35 & 1 \end{bmatrix} \Rightarrow \begin{bmatrix} 1 & 0 & 0 \\ -20/80 & 1 & 0 \\ -20/80 & -25/35 & 1 \end{bmatrix}$$

$$\Rightarrow \begin{bmatrix} 1 & 0 & 0 \\ -1/4 & 1 & 0 \\ -1/4 & -5/7 & 1 \end{bmatrix}.$$

To check the calculation, compute $\mathbf{LU}$ and compare it element by element with $\mathbf{A}$.

Remark: In order to use direct methods, we must ensure that the elements in the main diagonal are nonzero. If in the process of elimination, we end up with a zero element in the main diagonal, the order of equations need to be interchanged (switching either rows or columns, which is called pivoting).

In the literature, there are three **LU**-decomposition methods:

- Doolittle: where the elements of the main diagonal of lower triangular matrix are set to unity ($\mathbf{L}_{ii} = 1$).
- Crout: where the elements of the main diagonal of the upper triangular matrix are set to unity ($\mathbf{U}_{ii} = 1$).
- Choleski: where the lower triangular matrix is equal to the transpose of the upper triangular matrix ($\mathbf{L} = \mathbf{U}^T$). The Choleski decomposition is valid for symmetric matrices ($\mathbf{A} = \mathbf{LL}^T$) only.

2.5 Symmetric Matrix

In solving most engineering problems, such as solving elliptic (Laplace and Poisson equations) numerically, we end up with a symmetric matrix. Hence,

$$\mathbf{A} = \mathbf{LDL}^T. \tag{2.15}$$

Choleski decomposition is a special case of the above equation, where $\mathbf{D} = \mathbf{I}$.

GE process is used to generate an upper triangular matrix. Another symmetric matrix that arises in engineering and science problems is **penta-diagonal matrix**. All elements of the matrix are zero, except the main diagonal, two upper diagonal, and two lower diagonal elements are not zero, in general. For example,

$$\begin{bmatrix} d_1 & e_1 & f_1 & 0 & 0 & 0 & 0 \\ e_1 & d_2 & e_2 & f_2 & 0 & 0 & 0 \\ f_1 & e_2 & d_3 & e_3 & f_3 & 0 & 0 \\ 0 & f_2 & e_3 & d_4 & e_4 & f_4 & 0 \\ 0 & 0 & f_3 & e_4 & d_5 & e_5 & f_5 \\ 0 & 0 & 0 & f_4 & e_5 & d_6 & e_6 \\ 0 & 0 & 0 & 0 & f_5 & e_6 & d_7 \end{bmatrix}.$$

So, Choleski decomposition can be used with this matrix too.

Exercise: Develop an efficient algorithm to solve a system of algebraic equations ($\mathbf{Ax} = \mathbf{b}$), where the matrix $\mathbf{A}$ is a penta-diagonal matrix. *Hint*: Store only three vectors, $\mathbf{d}$, $\mathbf{e}$, and $\mathbf{f}$.

2.6   Iterative Methods

In most engineering and scientific problems, the resulting algebraic equations are sparse matrices (many elements are zero). Also, the direct methods are not efficient to solve a large number of equations. An exception may occur for dense matrices, for which direct methods might be preferable. Iterative methods are attractive to solve a large number of equations and are especially efficient in dealing with sparse matrices. The main drawback of the iterative methods is the convergence. There is no guarantee that the solution converges if the matrix is not strictly diagonally dominant. The main procedure of iterative method is to start with a guess solution and update the solution as

$$\mathbf{x}^{k+1} = \mathbf{x}^k + \alpha \mathbf{r}^k, \tag{2.16}$$

where $\mathbf{x}^{k+1}$, $\mathbf{x}^k$, α, and $\mathbf{r}^k$ are the updated values of the solution, previous value of the solution, a scalar multiplier, and residual, respectively. The method is called stationary if α is constant and does not change at each iteration (iteration independent); otherwise, the method is called non-stationary. The residual vector, $\mathbf{r}^k$, is

$$\mathbf{r} = \mathbf{b} - \mathbf{Ax}, \tag{2.17}$$

for the system of equations $\mathbf{Ax} = \mathbf{b}$.

2.7   Stationary Iterative Methods

Let us work on a simple example to illustrate a few concepts.

Example:

$$2x_1 + x_2 = 5, \tag{2.18}$$

$$-x_1 + 4x_2 = 6. \tag{2.19}$$

Of course, we can easily solve these equations, and the exact values of x_1 and x_2 are 1.5555 and 1.8888, respectively. However, let us solve them iteratively:

$$x_1 = \frac{1}{2}(5 - x_2), \tag{2.20}$$

$$x_2 = \frac{1}{4}(6 + x_1), \tag{2.21}$$

starting with initial guesses of $x_1^0 = 0$, $x_2^0 = 0$. The superscript notation represents the number of iterations.

Note from Table 2.1 the following: (1) The method is converging; (2) we are keeping old values of the variables, even though the updated values are available (in this example, x_1). This method is called Jacobi Method. Also, note that the matrix **A** is diagonally dominant:

$$\begin{bmatrix} 2 & 1 \\ -1 & 4 \end{bmatrix} \begin{bmatrix} x_1 \\ x_2 \end{bmatrix} = \begin{bmatrix} 5 \\ 6 \end{bmatrix}.$$

Let us slightly modify the method. Let us use the values of the variables as soon as they are available, see Table 2.2.

This method is called Gauss–Seidel method. It converges faster than Jacobi method.

Table 2.1: Successive solutions through iterations.

$x_1^0 = 0$	$x_2^0 = 0$
$x_1^1 = 2.50$	$x_2^1 = 1.50$
$x_1^2 = 1.75$	$x_2^2 = 2.125$
$x_1^3 = 1.4375$	$x_2^3 = 1.9375$
$x_1^4 = 1.5312$	$x_2^4 = 1.8585$
$x_1^5 = 1.5707$	$x_2^5 = 1.8927$
$x_1^6 = 1.5536$	$x_2^6 = 1.8926$

Table 2.2: Successive solutions through iterations using the most recent among them.

$x_1^0 = 0$	$x_2^0 = 0$
$x_1^1 = 2.50$	$x_2^1 = 2.125$
$x_1^2 = 1.4375$	$x_2^2 = 1.85937$
$x_1^3 = 1.5703$	$x_2^3 = 1.892578$
$x_1^4 = 1.55371$	$x_2^4 = 1.88842$
$x_1^5 = 1.555786$	$x_2^5 = 1.88894$
$x_1^6 = 1.55552$	$x_2^6 = 1.88888$

If we switch the rows as

$$\begin{bmatrix} -1 & 4 \\ 2 & 1 \end{bmatrix} \begin{bmatrix} x_1 \\ x_2 \end{bmatrix} = \begin{bmatrix} 6 \\ 5 \end{bmatrix},$$

the solution will diverge. Try to solve the equation iteratively using the same initial guesses and method.

The following is a generalization of the illustrated example.

The system of linear algebraic equations can formally be written as

$$\mathbf{0} = -\mathbf{A}\mathbf{x} + \mathbf{b}, \tag{2.22}$$

where $\mathbf{0}$ is a zero vector with the same size as $\mathbf{b}$. By adding and subtracting $\mathbf{I}\mathbf{x}$ from the above equation,

$$\mathbf{I}\mathbf{x} = (\mathbf{I} - \mathbf{A})\mathbf{x} + \mathbf{b}. \tag{2.23}$$

Iterative form of the above equation is

$$\mathbf{x}^{k+1} = (\mathbf{I} - \mathbf{A})\,\mathbf{x}^k + \mathbf{b}. \tag{2.24}$$

This class of methods is called stationary because the value of x_{i+1} depends only on x_i. As the iteration process proceeds, the history is lost.

Also, matrix $\mathbf{A}$ can be decomposed into the following sum:

$$\mathbf{A} = \mathbf{L} + \mathbf{D} + \mathbf{U}, \tag{2.25}$$

where $\mathbf{L}$, $\mathbf{D}$, and $\mathbf{U}$ are lower, diagonal, and upper matrices, respectively. For example, if $\mathbf{A}$ is $\mathbf{A} = \begin{bmatrix} 180 & -21 & -10 \\ -23 & 40 & -50 \\ -33 & -20 & 110 \end{bmatrix}$, then $\mathbf{L}$, $\mathbf{D}$, and $\mathbf{U}$ matrices are

$$\mathbf{L} = \begin{bmatrix} 0 & 0 & 0 \\ -23 & 0 & 0 \\ -33 & -20 & 0 \end{bmatrix}, \quad \mathbf{D} = \begin{bmatrix} 180 & 0 & 0 \\ 0 & 40 & 0 \\ 0 & 0 & 110 \end{bmatrix},$$

$$\mathbf{U} = \begin{bmatrix} 0 & -21 & -10 \\ 0 & 0 & -50 \\ 0 & 0 & 0 \end{bmatrix}.$$

2.7.1 Jacobi method

A system of linear algebraic equations can be written using index notation as

$$\sum_{j=1}^{n} a_{ij}x_j = b_i \quad i = 1, 2, 3, \ldots, n. \tag{2.26}$$

Rearranging the equation by moving all terms to the right-hand side except the element x_i, whose value we are interested to find,

$$x_i = \frac{1}{a_{ii}} \left(b_i - \sum_{j=1}^{i-1} a_{ij}x_j - \sum_{j=i+1}^{n} a_{ij}x_j \right) \quad i = 1, 2, 3, \ldots, n. \tag{2.27}$$

To make the above equation compact, add and subtract the term $a_{ii}x_i$ inside the right-hand side bracket, hence

$$x_i^{k+1} = x_i^k + \frac{1}{a_{ii}} \left(b_i - \sum_{j=1}^{n} a_{ij}x_j^k \right) \quad i = 1, 2, 3, \ldots, n, \tag{2.28}$$

where superscripts k and $k+1$ stand for number of iterations. Note that the last term on the right-hand side represents the error term (residual), and as the solution converges, the term should converge to the machine error (very small value, order of 10^{-4} or less). The error associated for the initial guess value is,

$$R_i^k = \left(b_i - \sum_{j=1}^{n} a_{ij}x_j^k \right) \quad i = 1, 2, 3, \ldots, n. \tag{2.29}$$

The improved value of x_i is calculated as

$$x_i^{k+1} = x_i^k + \frac{R_i^k}{a_{ii}} \quad i = 1, 2, 3, \ldots, n. \tag{2.30}$$

Jacobi method can be cast in matrix form as

$$\mathbf{x}^{k+1} = \mathbf{D}^{-1} \left[\mathbf{b} - (\mathbf{L} + \mathbf{U})\mathbf{x}^k \right] = \mathbf{D}^{-1}\mathbf{b} - \mathbf{D}^{-1}(\mathbf{L} + \mathbf{U})\mathbf{x}^k. \tag{2.31}$$

Remarks:

- Jacobi method uses the guess values without updating any of those values, even though the new (the updated) values of x_i are available to be used. This is one of the drawbacks of the method, which is the reason why Jacobi method usually has slow convergence. To overcome this issue, a method called Gauss–Seidel (GS) method is used instead. This latter method is more popular than Jacobi method. However, Jacobi method can easily be used with parallel processor machines since solution provided by each equation is independent from the outcome of the other equations.
- The convergence of the method is related to the spectral radius of matrix $\mathbf{D}^{-1}(\mathbf{L} + \mathbf{U})$. If it is less than 1, then Jacobi method converges, otherwise it diverges.
- The right-hand side vector $\mathbf{b}$ has no effect on the convergence process.

2.7.2   Gauss–Seidel method

GS method is a simple modification of Jacobi method. In GS method, as the updated value of x_j^{k+1} ($j < i$) is available. It will be used immediately without waiting to solve for all values of x_i^{k+1} as in Jacobi method. Hence, as soon as x_i^{k+1} is computed, it replaces the value of x_i^k in the machine memory. The equations can be written therefore as

$$x_i^{k+1} = \frac{1}{a_{ii}} \left(b_i - \sum_{j=1}^{i-1} a_{ij} x_j^{k+1} - \sum_{j=i+1}^{n} a_{ij} x_j^k \right) \quad i = 1, 2, 3, \ldots, n,$$

$$(2.32)$$

with adding and subtracting term $a_{ii} x_j^k$ on the right-hand side:

$$x_i^{k+1} = x_i^k + \frac{R_i^k}{a_{ii}} \quad i = 1, 2, 3, \ldots, n, \qquad (2.33)$$

where

$$R_i^k = b_i - \sum_{j=1}^{i-1} a_{ij} x_j^{k+1} - \sum_{j=i}^{n} a_{ij} x_j^k \quad i = 1, 2, 3, \ldots, n. \qquad (2.34)$$

In the matrix form, the GS method can be cast as

$$\mathbf{x}^{k+1} = (\mathbf{D}+\mathbf{L})^{-1}(\mathbf{b}-\mathbf{U}\mathbf{x}^k) = (\mathbf{D}+\mathbf{L})^{-1}\mathbf{b} - (\mathbf{D}+\mathbf{L})^{-1}\mathbf{U}\mathbf{x}^k. \tag{2.35}$$

Similar to Jacobi method, GS method converges only if the spectral radius of matrix $(\mathbf{D}+\mathbf{L})^{-1}\mathbf{U}$ is less than 1.

2.7.3 Examples using Jacobi and Gauss–Seidel methods

In this section, we examine the convergence of these methods when solving different problems. For instance, if we consider the following problem:

$$\mathbf{A}_1\mathbf{x} = \mathbf{b}, \tag{2.36}$$

where

$$\mathbf{A}_1 = \begin{bmatrix} 3 & -2 & 1 \\ -5 & 10 & 2 \\ -1 & 1 & 2 \end{bmatrix}, \tag{2.37}$$

$$\mathbf{x} = (x_1,\ x_2,\ x_3)^t, \tag{2.38}$$

and

$$\mathbf{b} = (1,\ 1,\ 1)^t, \tag{2.39}$$

then applying Jacobi and GS methods leads to a convergent solution for both, see Fig. 2.2.

Let us use the same right-hand side vector and examine the solution for different matrices $\mathbf{A}$, i.e., matrix $\mathbf{A}_2$, $\mathbf{A}_3$, or $\mathbf{A}_4$, as

$$\mathbf{A}_2 = \begin{bmatrix} 1 & 5 & 1 \\ 3 & 1 & 5 \\ 5 & 1 & 2 \end{bmatrix}, \quad \mathbf{A}_3 = \begin{bmatrix} -3 & 3 & -6 \\ -4 & 7 & -8 \\ 5 & 7 & -9 \end{bmatrix}, \quad \text{and}$$

$$\mathbf{A}_4 = \begin{bmatrix} 1 & 1 & 3 \\ 2 & 2 & 2 \\ 3 & 6 & 4 \end{bmatrix}. \tag{2.40}$$

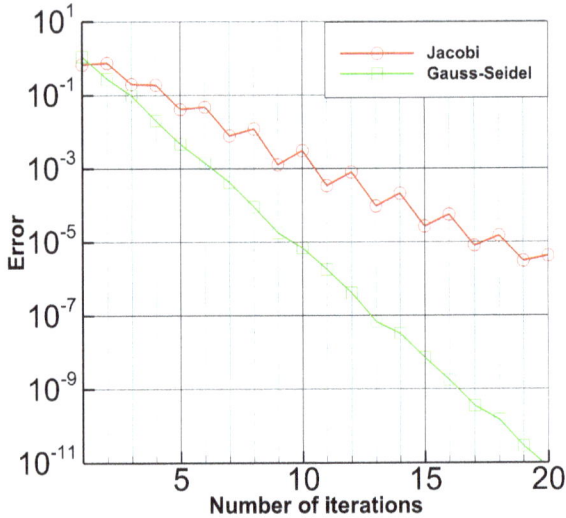

Figure 2.2:   Variations of the residual of Jacobi and Gauss–Seidel methods versus the iteration number when solving system (2.36) (semi-log scale).

Figure 2.3 shows the convergence of the solutions:

- For matrix A_2, both Jacobi and GS methods diverge (the error is growing exponentially).
- For matrix A_3, Jacobi method converges and GS method diverges.
- For matrix A_4, Jacobi method diverges and GS method converges.

The examination of the spectral radius of matrix $D^{-1}(L + U)$ and $(D + L)^{-1}U$ (see Table 2.3) associated with Jacobi and GS methods, respectively, explains the following trends: Spectral radius must be less than one for the method to converge, in which case the smaller the spectral radius, the faster the method converges.

2.7.4   Successive over-relaxation method

The residual term, R_i, in GS method, for example, should approach the machine zero as the solution converges. Hence, any multipliers of R_i does not affect the end results. The main idea of successive over-relaxation (SOR) is to multiply the residual term with a factor greater than one, which was suggested by Southwell in 1940.

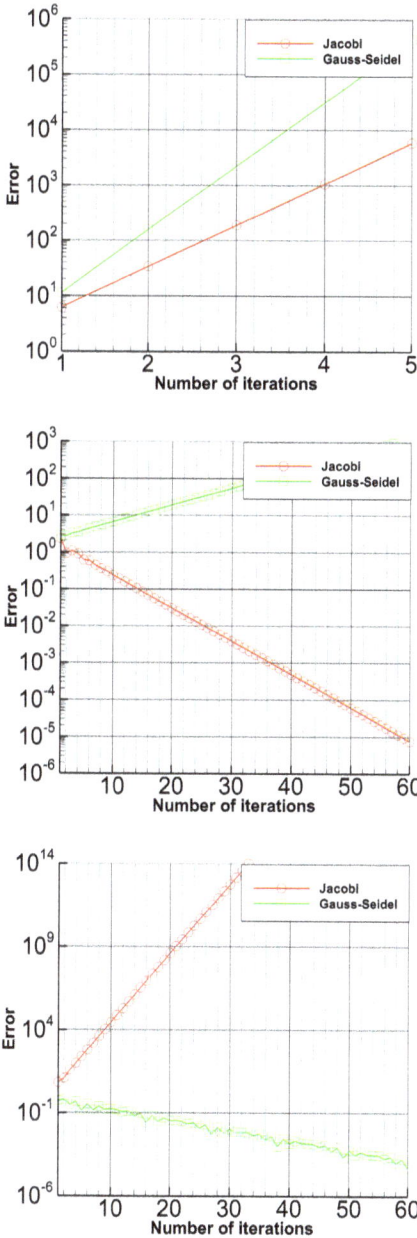

Figure 2.3: Variations of the residual, in semi-log scale, of Jacobi and Gauss–Seidel methods versus the iteration number using matrices $\mathbf{A}_2$ (top), $\mathbf{A}_3$ (middle), and $\mathbf{A}_4$ (bottom).

Table 2.3:   The spectral radius
of matrices $\mathbf{D}^{-1}(\mathbf{L} + \mathbf{U})$ and
$(\mathbf{D} + \mathbf{L})^{-1}\mathbf{U}$.

	Jacobi	Gauss–Seidel
$\mathbf{A}_1$	0.545194	0.258199
$\mathbf{A}_2$	5.60518	14.0523
$\mathbf{A}_3$	0.813309	1.11111
$\mathbf{A}_4$	1	0.353553

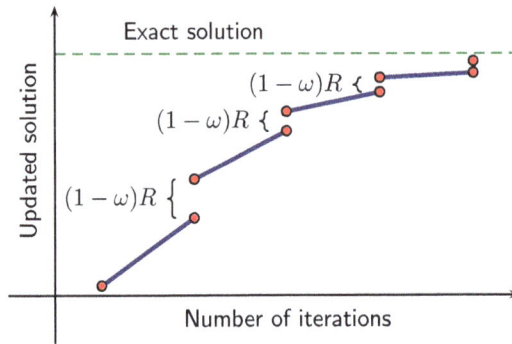

Figure 2.4:   Illustration of the convergence speed up using SOR.

The concept of SOR is illustrated in Fig. 2.4. Let us denote the factor by ω. Hence,

$$x_i^{k+1} = x_i^k + \frac{\omega R_i^k}{a_{ii}}, \tag{2.41}$$

where

$$R_i^k = b_i - \sum_{j=1}^{i-1} a_{ij}x_j^{k+1} - \sum_{j=i}^{n} a_{ij}x_j^k \quad i = 1, 2, 3, \ldots, n. \tag{2.42}$$

For SOR, $\omega > 1$. If the iterative method already converges, such a small modification accelerates the solution drastically. The question is what is the optimal value of ω? Practically, the ω values should lie between 1 and 2. As the number of equations increases, selecting a higher value of ω is recommended. It may be worth using different values of ω, less than 2, and measure the number of iterations needed to achieve the required solution, if we are solving the same kind of equations many times. However, starting with value of 1.7 for a large system of equations (order of 1000 or 10000) may be a good strategy.

If the system of equations is not diagonally dominant, using $w > 1$ may lead to divergence of the solution. In such a case, it is recommended to use $0 < w < 1$, which is called under-relaxation. However, the convergence of the solution degrades significantly.

2.8 Non-stationary Methods

Jacobi, GS, and SOR methods are considered stationary methods. Those methods have the following form:

$$\mathbf{x}^{k+1} = \mathbf{c}\mathbf{x}^k + \mathbf{e}, \tag{2.43}$$

for a linear system of equations of $\mathbf{Ax} = \mathbf{b}$. In the above equation, $\mathbf{c}$ and $\mathbf{e}$ are independent of the iteration counter k. On the other hand, if they are changing with the iteration, the method is qualified as **non-stationary**. In the literature, many methods were introduced, such as **steepest descent, conjugate gradient, minimum residual and symmetric**, and **bi-conjugate gradient**. We refer the reader to the text *Iterative Methods for Sparse Linear Systems, 2nd edition* by Yousef Saad (2000) for more details. The topic of non-stationary methods is outside the focus of this text. As we mentioned before, the interested reader is encouraged to consult the book by Saad. In the following section, the algorithm of applying the method is illustrated.

2.8.1 Steepest descent method

Before discussing steepest descent method (SDM), we need to understand the maxima and minima of the multidimensional quadratic equations. A one-dimensional general quadratic algebraic equation can be written as

$$f(x) = ax^2 + bx + c. \tag{2.44}$$

To find a minimum or a maximum of the equation, set the derivative of function $f(x)$ to zero:

$$f(x)' = 2ax + b = 0. \tag{2.45}$$

The point $x = \frac{-b}{2a}$ is a minimum or a maximum. If the second derivative of the function $(f'' = 2a)$ is positive $(a > 0)$, then x represents
the minimum value of the function. If the second derivative is negative $(a < 0)$, then x represents the maximum value of the function.

Example: A household has a chain (wire) of length 20 m. The resi-
dent of the household would like to use that chain to make a vegetable
garden of a rectangular shape (L and W are length and width of the
area, respectively) that surrounds the garden with the maximum area
possible. The area, A, of the rectangle is

$$A = WL, \qquad (2.46)$$

and the perimeter of the rectangle is $2L + 2W = 20$, hence

$$L = 10 - W. \qquad (2.47)$$

The area can be reformulated as

$$A = W(10 - W) = 10W - W^2. \qquad (2.48)$$

Differentiate A with respect of W and set the result to zero:

$$\frac{dA}{dW} = 10 - 2W = 0. \qquad (2.49)$$

Hence, $W = 5$. The second derivative, $\frac{d^2A}{dW^2} = -10$, which is negative.
Hence, the area with $W = 5$ and $L = 5$ represents a square.
   A multidimensional quadratic equation can be written in
matrix form as

$$f(x_1, x_2, x_3, \ldots, x_n) = \tfrac{1}{2}\mathbf{x}^T \mathbf{A}\mathbf{x} - \mathbf{b}^T \mathbf{x} + c, \qquad (2.50)$$

where $\mathbf{x}$ is a column vector, $\mathbf{x}^T = (x_1, x_2, x_3, \ldots, x_n)$, $\mathbf{A}$ a $n \times n$
matrix, $\mathbf{b}$ a column vector, $\mathbf{b}^T = (b_1, b_2, b_3, \ldots, b_n)$, and c a scalar.

By partial differentiation of Eq. (2.50), we retrieve

$$f'(x_1, x_2, x_3, \ldots, x_n) = \nabla f = \mathbf{A}\mathbf{x} - \mathbf{b}, \qquad (2.51)$$

where

$$(\nabla f)^T = \left(\frac{\partial f}{\partial x_1}, \frac{\partial f}{\partial x_2}, \ldots, \frac{\partial f}{\partial x_n} \right). \qquad (2.52)$$

Let us introduce the following definition for error vector, $\mathbf{e}$:

$$\mathbf{e}^k = \mathbf{x}^k - \mathbf{x}, \qquad (2.53)$$

which indicates how far from the exact solution we are. For the given vector $\mathbf{x}$, the residual is

$$\mathbf{r}^k = \mathbf{b} - \mathbf{A}\mathbf{x}^k. \qquad (2.54)$$

Hence,

$$\mathbf{r}^k = \mathbf{A}\mathbf{e}^k. \qquad (2.55)$$

Also, from Eqs. (2.51) and (2.54),

$$\mathbf{r}^k = -f'(x_1, x_2, x_3, \ldots, x_n). \qquad (2.56)$$

Assume that the first correct step to get $\mathbf{x}^1$ is such that

$$\mathbf{x}^1 = \mathbf{x}^0 + \alpha \mathbf{r}^0, \qquad (2.57)$$

where α is an unknown factor. In general,

$$\frac{d\mathbf{x}}{d\alpha} = \mathbf{r}. \qquad (2.58)$$

Also, chain rule gives

$$\frac{df}{d\alpha} = \frac{df}{d\mathbf{x}} \cdot \frac{d\mathbf{x}}{d\alpha}. \qquad (2.59)$$

Example: Consider a two-dimensional vector $\mathbf{X}^T = (x_1, \ x_2)$, a symmetric matrix $\mathbf{A}$:

$$\mathbf{A} = \begin{bmatrix} 2.5 & 1.2 \\ 1.2 & 4 \end{bmatrix}, \quad \text{and} \quad \mathbf{b} = \begin{bmatrix} 2 \\ 5 \end{bmatrix}.$$

Hence,

$$f(x_1, x_2) = 1.25x_1^2 + 1.2x_1x_2 + 2x_2^2 - 2x_1 - 5x_2 + c, \qquad (2.60)$$

where c is an arbitrarily constant. The derivatives of $f(x_1, x_2)$ with respect of x_1 and x_2 are

$$\frac{\partial f}{\partial x_1} = 2.5x_1 + 1.2x_2 - 2 \qquad (2.61)$$

and

$$\frac{\partial f}{\partial x_2} = 1.2x_1 + 4x_2 - 5, \qquad (2.62)$$

respectively. Set those derivatives to zero and solve for x_1 and x_2, which yields ($x_1 = 0.2336$, $x_2 = 1.18$). Note that we are solving the following set of equations: $\mathbf{Ax} = \mathbf{b}$. Whether the values of x_1 and x_2 correspond to the minimum of function f depends on the positivity of the second derivatives of f with respect to x_1 and x_2, which in this case can be easily verified. Hence, the solution gives the minimum of function f.

Coming back to the general case, the gradient $(-\nabla f)$ of function f gives the direction (vector) along which the function is decreasing rapidly. Hence,

$$-\nabla f(x_{i+1}) = b - Ax_{i+1} = r_{i+1}. \qquad (2.63)$$

(Of course, if our estimate for $\mathbf{x}$ is correct, the gradient should be zero.) However, the gradient only gives the direction we need to follow to reach the minimum straightforwardly. However, we do not know the step length. Therefore, we need to multiply it by a scalar, say α, to adjust the vector length.

To find out the solution for the next iteration, the function needs to be followed in the gradient direction as

$$\mathbf{x}^{k+1} = \mathbf{x}^k - \alpha^k \nabla \mathbf{f}^k = \mathbf{x}^k + \alpha^k \mathbf{r}^k \left. \frac{\partial f}{\partial \alpha} \right|^{k+1} = 0. \qquad (2.64)$$

Since f is function of $x_1, x_2, x_3, \ldots, x_n$),

$$\left(\frac{\partial f}{\partial x_1}\frac{\partial x_1}{\partial \alpha_1} + \frac{\partial f}{\partial x_2}\frac{\partial x_2}{\partial \alpha_2} + \frac{\partial f}{\partial x_3}\frac{\partial x_3}{\partial \alpha_3} + \cdots + \frac{\partial f}{\partial x_n}\frac{\partial x_n}{\partial \alpha_n} \right)_{k+1} = 0. \qquad (2.65)$$

From Eq. (2.58), $\partial x_i / \partial \alpha = r_i$. Hence,

$$\left(\frac{\partial f}{\partial x_1} r_1 + \frac{\partial f}{\partial x_2} r_2 + \frac{\partial f}{\partial x_3} r_3 + \cdots + \frac{\partial f}{\partial x_n} r_n \right)_{k+1} = 0. \qquad (2.66)$$

In matrix form, $\nabla \mathbf{f}_{i+1}^T \mathbf{r}_i = 0$, which leads to $-\mathbf{r}_{i+1}^T \mathbf{r}_i = 0$. Hence, $\mathbf{r}_{i+1}^T$ is orthogonal to $\mathbf{r}_i$. Note that

$$\mathbf{r}_{i+1}^T \mathbf{r}_i = (\mathbf{b} - \mathbf{A} \mathbf{x}_{i+1})^T \mathbf{r}_i = 0. \qquad (2.67)$$

Since $\mathbf{x}_{i+1} = \mathbf{x}_i + \alpha \mathbf{r}_i$, then

$$(\mathbf{b} - \mathbf{A}(\mathbf{x}_i + \alpha \mathbf{r}_i))^T \mathbf{r}_i = 0. \qquad (2.68)$$

Expanding, we get

$$(\mathbf{b} - \mathbf{A} \mathbf{x}_i - \alpha \mathbf{A} \mathbf{r}_i)^T \mathbf{r}_i = 0. \qquad (2.69)$$

Note also that $\mathbf{b} - \mathbf{A} \mathbf{x}_i$ equals to $\mathbf{r}_i$. Hence,

$$\alpha = \frac{\mathbf{r}_i^T \mathbf{r}_i}{(\mathbf{A} \mathbf{r}_i)^T \mathbf{r}_i}. \qquad (2.70)$$

For a symmetric matrix, i.e., $\mathbf{A} = \mathbf{A}^T$, the above equation can be written as

$$\alpha = \frac{\mathbf{r}_i^T \mathbf{r}_i}{\mathbf{r}_i^T \mathbf{A} \mathbf{r}_i}. \qquad (2.71)$$

The algorithm of steepest descent method is as follows:

(1) Guess values of $\mathbf{x}$, $x_1, x_2, x_3, \ldots$.
(2) Calculate $\mathbf{r}_i = \mathbf{b} - \mathbf{A} \mathbf{x}$.
(3) Evaluate α, see Eq. (2.70).
(4) Update $\mathbf{x}$, $\mathbf{x}_{i+1} = \mathbf{x}_i + \alpha \mathbf{r}_i$.
(5) If $\|\mathbf{x}^{k+1} - \mathbf{x}^k\|$ is above a given tolerance, then go to step (2), else stop.

End of the algorithm.

Example: Solve the following set of linear algebraic equations using SDM. Start the solution with $x_1 = x_2 = x_3 = 1$:

$$\begin{bmatrix} 4 & 1 & 0 \\ 1 & 4 & 1 \\ 0 & 1 & 4 \end{bmatrix} \begin{bmatrix} x_1 \\ x_2 \\ x_3 \end{bmatrix} = \begin{bmatrix} 2 \\ 1 \\ -1 \end{bmatrix}.$$

Hence,
$\mathbf{r}_1^T = (-3, -5, -6)$, $\alpha = 0.189$, then updated $\mathbf{x}$ is

$$[x_1 \quad x_2 \quad x_3] = [0.432, \ 0.054, \ -0.135]. \tag{2.72}$$

In the second iteration, $\mathbf{r}_2^T = (0.216, \ 0.486, \ -0.513)$, $\alpha = 0.28807$. The update values of $\mathbf{x}$ are

$$(x_1, \ x_2, \ x_3) = (0.4947, \ 0.1942, \ -0.2830). \tag{2.73}$$

In the third iteration, $\mathbf{r}_3^T = (-0.1730, \ 0.0115, \ -0.0619)$, $\alpha = 0.26041$. The update values of $\mathbf{x}$ are

$$(x_1, \ x_2, \ x_3) = (0.4496, \ 0.1972, \ -0.2991). \tag{2.74}$$

Hence, $\mathbf{r}_4^T = (0.00420, \ 0.0607, \ -0.00004)$, $\alpha = 0.24249$. The updated values of $\mathbf{x}$ are

$$\mathbf{x} = (0.4506, \ 0.2119, \ -0.2992)^T. \tag{2.75}$$

2.8.2 Conjugate gradient method

This method applied when solving a system of linear algebraic equations ($\mathbf{Ax} = \mathbf{b}$). The matrix $\mathbf{A}$ is symmetric positive definite. As mentioned before, a symmetric matrix is such that $\mathbf{A} = \mathbf{A}^T$ and a positive definite matrix is such that all of its eigenvalues are positive. In general, the diagonally dominate, symmetric matrices are a sub-class of symmetric positive definite matrices. In conjugate gradient (CG) method, the updated values of vector $\mathbf{x}$ are

$$\mathbf{x}^{k+1} = \mathbf{x}^k + \alpha^k \mathbf{P}^k, \tag{2.76}$$

where $\mathbf{P}$ is a vector in the search direction and α is a scalar that determines the step length. Note that $\mathbf{P}$ and α are not constants for the cases of the stationary methods. As we discussed in the previous section, to find the steepest direction, we showed that the negative gradient of the quadratic function is the direction to follow.

In the previous section, the error of the previous iteration, $\mathbf{r}^k$, is orthogonal to the updated error, $\mathbf{r}^{k+1}$, $(\mathbf{r}^{k+1^T}\mathbf{r}^k)$, see the statement after Eq. (2.66). Therefore, in SDM, we are following stair steps to reach the solution. The idea of CG method is to find a right direction. At this point, we need to introduce two kinds of errors. One of those is

$$\mathbf{r}^k = \mathbf{b} - \mathbf{A}\mathbf{x}^k, \tag{2.77}$$

which was already introduced in the previous sections. Another error is the measure of how far we are from the exact solution, i.e.,

$$\mathbf{e}^k = \mathbf{x}^k - \mathbf{x}. \tag{2.78}$$

The relation between those errors is

$$\mathbf{r}^k = -\mathbf{A}\mathbf{e}^k, \tag{2.79}$$

The problem is that we do not know $\mathbf{e}^k$; otherwise, we would know the solution! The main idea is to find a vector, say $\mathbf{d}^k$ $\mathbf{A}$, that is orthogonal to $\mathbf{e}^{k+1}$, i.e.,

$$\mathbf{d}^{k^T}\mathbf{A}\mathbf{e}^{k+1} = 0. \tag{2.80}$$

Since $\mathbf{e}^{k+1} = \mathbf{e}^k + \alpha^k\mathbf{d}^k$, hence

$$\mathbf{d}^{k^T}\mathbf{A}(\mathbf{e}^k + \alpha^k\mathbf{d}^k) = 0. \tag{2.81}$$

Then,

$$\alpha^k = -\frac{\mathbf{d}^{k^T}\mathbf{A}\mathbf{e}^k}{\mathbf{d}^{k^T}\mathbf{A}\mathbf{d}^k}. \tag{2.82}$$

Let $\mathbf{M}$ be a matrix of the same size as matrix $\mathbf{A}$. $\mathbf{M}$ is called pre-conditioner. For a non-preconditioned problem, $\mathbf{M} = \mathbf{I}$.

The CG algorithm is as follows:

$$\mathbf{r}^0 = \mathbf{B} - \mathbf{A}\mathbf{X}^0,$$

$$\mathbf{d}^0 = \mathbf{M}^{-1}\mathbf{r}^0,$$

$$\alpha^k = \frac{\mathbf{r}^{k^T}\mathbf{M}^{-1}\mathbf{r}^k}{\mathbf{d}^{k^T}\mathbf{A}\mathbf{d}^k},$$

$$\mathbf{x}^{k+1} = \mathbf{x}^k + \alpha^k\mathbf{d}^k,$$

$$\mathbf{r}^{k+1} = \mathbf{r}^k - \alpha^k\mathbf{A}\mathbf{d}^k,$$

$$\beta^{k+1} = \frac{\mathbf{r}^{k+1^T}\mathbf{M}^{-1}\mathbf{r}^{k+1}}{\mathbf{r}^{k^T}\mathbf{M}^{-1}\mathbf{r}^k},$$

$$\mathbf{d}^{k+1} = \mathbf{M}^{-1}\mathbf{r}^{k+1} + \beta^{k+1}\mathbf{d}^k.$$

End of the algorithm.

Example: Let us consider the problem $\mathbf{A}\mathbf{x} = \mathbf{b}$, where

$$\mathbf{A} = \begin{bmatrix} 0.2 & 0.1 & 1 & 1 & 0 \\ 0.1 & 4 & -1 & 1 & -1 \\ 1 & -1 & 60 & 0 & -2 \\ 1 & 1 & 0 & 8 & 4 \\ 0 & -1 & 2 & 4 & 700 \end{bmatrix} \quad \text{and} \quad \mathbf{b} = \begin{bmatrix} 1 \\ 2 \\ 3 \\ 4 \\ 5 \end{bmatrix}.$$

We choose the pre-conditioning matrix $\mathbf{M}$ such that

$$\mathbf{M}^{-1} = \mathbf{D}^{-1/2} = \begin{bmatrix} \frac{1}{\sqrt{0.2}} & 0 & 0 \ 0 & 0 \\ 0 & \frac{1}{\sqrt{4}} & 0 & 0 & 0 \\ 0 & 0 & \frac{1}{\sqrt{60}} & 0 & 0 \\ 0 & 0 & 0 & \frac{1}{\sqrt{8}} & 0 \\ 0 & 0 & 0 & 0 & \frac{1}{\sqrt{700}} \end{bmatrix}.$$

Table 2.4: Comparison of the convergence rate of different iterative methods.

Method	Number of iterations	$\|\mathbf{x}^{\text{last iter.}} - \mathbf{x}^{\text{before last iter.}}\|$
Jacobi	49	0.00305834
Gauss–Seidel	15	0.02445559
SOR ($\omega = 1.25$)	7	0.00818607
CG (without preconditioning)	5	0.00629785
Preconditioned CG ($\mathbf{M}^{-1} = \mathbf{D}^{-1/2}$)	4	0.00009312

The convergence of Jacobi, GS, SOR, and CG methods applied to this system is shown in Table 2.4.

2.9 Problems

Problem 1: Develop a computer code to solve n linear algebraic equations using (a) GE method, (b) GS iteration method, and (c) SOR method with different values of ω. Test the methods by solving the following system of equations:

$$\begin{bmatrix} 10 & -1 & 4 & 0 & 2 & 9 \\ 0 & 25 & -2 & 7 & 8 & 4 \\ 1 & 0 & 15 & 7 & 3 & -2 \\ 6 & -1 & 2 & 23 & 0 & 8 \\ -4 & 2 & 0 & 5 & -25 & 3 \\ 0 & 7 & -1 & 5 & 4 & -22 \end{bmatrix} \begin{bmatrix} x_1 \\ x_2 \\ x_3 \\ x_4 \\ x_5 \\ x_6 \end{bmatrix} = \begin{bmatrix} 19 \\ 2 \\ 13 \\ -7 \\ -9 \\ 2 \end{bmatrix}.$$

Also, modify the above equation into a tri-diagonal matrix and use Thomas algorithm iteratively.

Hint: To convert the above matrix into a tri-diagonal matrix, move all the elements not on the diagonal, upper diagonal, and lower diagonal to the right-hand side and modify the vector **b**. But since those elements are unknown, they need to be guessed. Hence,

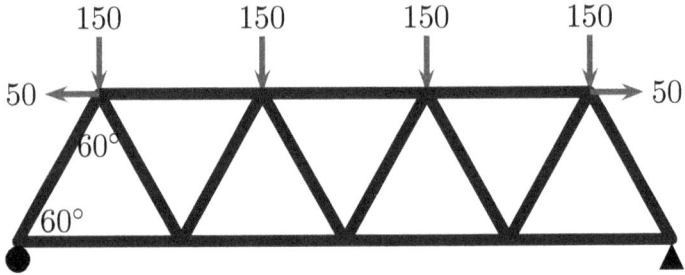

Figure 2.5:   Illustration of the structure and loads applied on it.

$$
\begin{bmatrix}
10 & -1 & 0 & 0 & 0 & 0 \\
0 & 25 & -2 & 0 & 0 & 0 \\
0 & 0 & 15 & 7 & 0 & 0 \\
0 & 0 & 2 & 23 & 0 & 0 \\
0 & 0 & 0 & 5 & -25 & 3 \\
0 & 0 & 0 & 0 & 4 & -22
\end{bmatrix}
\begin{bmatrix}
x_1 \\ x_2 \\ x_3 \\ x_4 \\ x_5 \\ x_6
\end{bmatrix}
=
\begin{bmatrix}
19 - 4x_3 - 2x_5 - 9x_6 \\
2 - 7x_4 - 8x_5 - 4x_6 \\
13 - x_1 - 3x_5 + 2x_6 \\
-7 - 6x_1 + x_2 - 8x_6 \\
-9 + 4x_1 - 2x_2 \\
2 - 7x_2 + x_3 - 5x_4
\end{bmatrix}.
$$

Problem 2: Consider a frame structure, shown in Fig. 2.5, made of eight elements and loaded at specific points with 150 force units and two tensions of 50 force units each. Estimate the tension and compression forces applied to each element and forces at the supporting points.

Problem 3: Solve the following system of equations using LU decomposition, then using CG with and without pre-conditioning:

$$
\begin{bmatrix}
10 & 2 & 6 & 0 \\
1 & 4 & 1 & 1 \\
-1 & 2 & 6 & -2 \\
2 & -2 & 2 & 8
\end{bmatrix}
\begin{bmatrix}
x_1 \\ x_2 \\ x_3 \\ x_4
\end{bmatrix}
=
\begin{bmatrix}
32 \\ 11 \\ 23 \\ -4
\end{bmatrix}.
$$

Extra Reading

W.K. Knapp. *Basic Algebra*. Cornerstones, Birkhäuser, Boston, MA, 2006.

Chapter 3

Interpolation and Fitting

3.1 Background

Experimental data are usually discrete points. For example, consider the investigation of pressure drop in a duct as a function of velocity or other parameters: The pressure is measured for a set of velocity values. Interpolation is done to find the value of pressure for arbitrary velocity values not measured but within the range of the measured velocities. Moreover, the tabulated data can be converted into an equation, which can be used for computer programming. Table 3.1 summarizes the data points collected from a given experimental setup. How can we proceed if we need to estimate the pressure at a velocity of 2.25 m/s, a value not measured?

We can use either interpolation or fitting.

Note that in order to perform an interpolation, the data in the example are sampled with uniform interval of 0.5 m/s. It is not necessary for the interval be the same. However, interpolating data at uniform intervals is much easier to manipulate than data at non-uniform intervals, although some interpolation methods, such as Chebyshev interpolation, do require non-uniform (but specific) sampling data. Nowadays, machine learning algorithms use interpolation methods to predict certain outputs. Furthermore, interpolation is heavily used in image processing, computer graphics, and computer security algorithms.

Fitting by regression is used to establish a **smooth** correlation between dependent variables and independent variables. In most experimental measurements, the collected data are prone to errors

Table 3.1:   Some data used for interpolation.

Velocity (m/s)	Pressure (Pa)
1.0	200.5
1.5	182.0
2.0	176.2
2.5	168.5
3.0	160.1
3.5	155.8
4.0	149.8

and uncertainty. Also, in statistical analyses, the data are essentially scattered. A best curve fitting is needed to correlate those data. Therefore, finding the best fit is a much better and more useful strategy than interpolation.

Note: Except perhaps for machine learning algorithms, it is usually important to understand the physics of the problem before fitting a data in establishing a correlation for the problem. Hence, the developed formula or correlation reflects the physics of the problem. For instance, if we know that the dependent variable exponentially decreases with the independent variable, then it is wise to fit with exponential functions in the regression.

3.2   Linear Interpolation

Linear interpolation is passing a line between two data points (called nodes). Therefore, the function shows discontinuity of higher-order derivatives at each point. The piecewise function between any two points is

$$f(x) = a + bx, \tag{3.1}$$

where a and b are constants, which can be evaluated by setting the value of the function at two different points, say x_i and x_{i+1}, i.e.,

$$b = \frac{f(x_{i+1}) - f(x_i)}{x_{i+1} - x_i} \quad \text{and} \quad a = f(x_i) - bx_i. \tag{3.2}$$

However, higher-order interpolations are more common but have some issues, which will be discussed later. In general, an $n - 1$ polynomial needs n points (nodes). For instance, a second-order polynomial requires three points. Popular methods for interpolations are Lagrange, Hermite, and spline interpolations.

3.3 Higher-Order Interpolation

A higher-order polynomial is, for instance,

$$P(x) = a_0 + a_1 x + a_2 x^2 + a_3 x^3. \tag{3.3}$$

The four unknown coefficients (a_0, a_1, a_2, and a_3) need four nodes (x, y), say $(2, 20.0)$, $(3.5, 31.9)$, $(4.7, 34.1)$, and $(6.7, 36.5)$. Hence,

$$\begin{bmatrix} 1 & 2 & 2^2 & 2^3 \\ 1 & 3.5 & 3.5^2 & 3.5^3 \\ 1 & 4.7 & 4.7^2 & 4.7^3 \\ 1 & 6.7 & 6.7^2 & 6.7^3 \end{bmatrix} \begin{bmatrix} a_0 \\ a_1 \\ a_2 \\ a_3 \end{bmatrix} = \begin{bmatrix} 20 \\ 31.9 \\ 34.1 \\ 36.5 \end{bmatrix}.$$

The unknown coefficients can be determined by solving the system of linear equations. However, **higher-order polynomials are not recommended for a few reasons.** Usually, they lead to an ill-conditioned matrix. (To mitigate the matrix ill-conditioning issue, one can interpolate semi-logarithm of positive nonzero independent set of data, that is $(\ln x, y)$, instead of (x, y), but this may reduce the sensitivity.) The polynomial that passes through all data points may result in an oscillatory pattern. Furthermore, adding new data points is not a straightforward procedure since the entire interpolation must be processed again.

3.4 Lagrange Interpolation

The $n - 1$ polynomial can be written as

$$f(x) = a_{n-1} x^{n-1} + a_{n-2} x^{n-2} + a_{n-3} x^{n-3} + \cdots + a_2 x^2 + a_1 x + a_0. \tag{3.4}$$

Evaluating the above function at each point, say (x_i, y_i), yields a system of linear algebraic equations:

$$a_{n-1}x_i^{n-1} + a_{n-2}x_i^{n-2} + a_{n-3}x_i^{n-3} + \cdots + a_2x_i^2 + a_1x_i + a_0 = y_i$$

$$i = 1, \ldots, n. \tag{3.5}$$

However, the mentioned approach is not recommended because for large n, the matrix may become ill-conditioned and difficult to solve. To overcome this issue, Lagrange form of a polynomial is recommended. Lagrange polynomial is

$$f(x) = \sum_{i=0}^{n} L_i(x)f(x_i), \tag{3.6}$$

where

$$L_i = \prod_{\substack{j=0 \\ j \neq i}}^{n} \frac{x - x_j}{x_i - x_j}. \tag{3.7}$$

Note that L_i has zero values at all x values except at the location of interest, namely x_i, at which it equals unity (one). For example, a second-order polynomial can be written as

$$f(x) = \frac{(x - x_1)(x - x_2)}{(x_0 - x_1)(x_0 - x_2)}f(x_0) + \frac{(x - x_0)(x - x_2)}{(x_1 - x_0)(x_1 - x_2)}f(x_1)$$

$$+ \frac{(x - x_0)(x - x_1)}{(x_2 - x_0)(x_2 - x_1)}f(x_2). \tag{3.8}$$

Lagrange polynomial can be derived using Taylor series, then approximating the derivatives by using finite differences. For example, the first-order Taylor series for a function, $f(x)$, expanded at $x = x_0$ is

$$f(x) = f(x_0) + \frac{df}{dx}\bigg|_{x_0}(x - x_0). \tag{3.9}$$

The function can be evaluated at $x = x_1$:

$$f(x_1) = f(x_0) + \frac{df}{dx}\bigg|_{x_0}(x_1 - x_0). \tag{3.10}$$

Hence,

$$\frac{df}{dx}\bigg|_{x_0} = \frac{f(x_1) - f(x_0)}{x_1 - x_0}. \tag{3.11}$$

By substituting into Eq. (3.10),

$$f(x) = f(x_0) + \frac{f(x_1) - f(x_0)}{x_1 - x_0}(x - x_0) \tag{3.12}$$

and rearranging gives

$$f(x) = \frac{x - x_1}{x_0 - x_1}f(x_0) + \frac{x - x_0}{x_1 - x_0}f(x_1), \tag{3.13}$$

which is the Lagrange first-order polynomial.

In a similar way, we can derive second-order Lagrange polynomial. The second-order Taylor series is

$$f(x) = f(x_0) + \frac{df}{dx}(x - x_0) + \frac{1}{2}\frac{d^2 f}{dx^2}(x - x_0)^2. \tag{3.14}$$

$\frac{df}{dx}$ can be approximated by using forward difference as $\frac{f(x_1) - f(x_0)}{x_1 - x_0}$ for $\frac{df}{dx}$, and $\frac{d^2 f}{dx^2}$ can be approximated by using central difference as

$$\frac{d^2 f}{dx^2} = \left[\frac{f(x_2) - f(x_1)}{x_2 - x_1} - \frac{f(x_1) - f(x_0)}{x_1 - x_0}\right]\frac{1}{(x_2 - x_0)}, \tag{3.15}$$

which can be rewritten as

$$\frac{d^2 f}{dx^2} = \frac{f(x_2)}{(x_2 - x_0)(x_2 - x_1)} + \frac{f(x_1)}{(x_1 - x_0)(x_1 - x_2)}$$
$$+ \frac{f(x_0)}{(x_0 - x_1)(x_0 - x_2)}. \tag{3.16}$$

Hence, by substituting the above differences in Eq. (3.14) leads to Eq. (3.8).

Example: Approximate $y = e^x$. For values of $x = 1$, 2, and 3, we have $y = 2.718$, 7.389, and 20.085, respectively. Estimate the value of y at $x = 1.5$ and 2.5.

$$y = f(x) = \frac{(x-2)(x-3)}{(1-2)(1-3)}2.718 + \frac{(x-1)(x-3)}{(2-1)(2-3)}7.389$$

$$+ \frac{(x-1)(x-2)}{(3-1)(3-2)}20.085, \tag{3.17}$$

$$y = 1.359(x-2)(x-3) - 7.389(x-1)(x-3)$$

$$+ 10.1042(x-1)(x-2). \tag{3.18}$$

At $x = 1.5$, $y = 1.01925 + 5.54175 - 2.526 = 4.03495$, while the exact value is 4.4817. The error is about 10%. For $x = 2.5$, $y = -0.33975 + 5.54175 + 7.576 = 12.778$, while the exact value is 12.1825. The error is about 4.88%.

3.5   Chebyshev Polynomial

As it is clear from the previous sections, we are free to select the sampling (independent variable) data points. However, the error between the fitted polynomial and the function is not uniformly distributed. In general, the polynomial leads to significant error at the end points compared with the interior points. Had we the freedom to select the sampling points, it is preferable to cluster more points at the ends of the domain. Chebyshev method achieves that goal by requiring a specific distribution of the independent data points in order to ensure an evenly distributed error.

Chebyshev polynomial (CP) was introduced by the Russian mathematician, Lvovich Chebyshev (1821–1894). The CP is valid for the range of -1 to 1 because it is based on cos functions. However, the interval from a to b can be linearly transformed into -1 to 1 as follows.

Variable t, such that $a \leqslant t \leqslant b$, needs to be transformed into another variable, say x, where $-1 \leqslant x \leqslant 1$. Let us assume that

$$t = c_0 + c_1 x, \tag{3.19}$$

where c_0 and c_1 are unknown constants that need to be evaluated. For $t = a$, $x = -1$. Hence,

$$a = c_0 - c_1, \tag{3.20}$$

and for $t = b$, $x = 1$. Hence,

$$b = c_0 + c_1. \tag{3.21}$$

Therefore, $c_0 = \frac{a+b}{2}$ and $c_1 = \frac{b-a}{2}$, and we can write

$$t = \frac{a+b}{2} + \frac{b-a}{2}x. \tag{3.22}$$

The first kind of CP is defined as

$$T_n(x) = \cos(n\theta), \quad \theta = \text{arc } \cos(x). \tag{3.23}$$

For $n = 1$, $\cos 0 = 1$, and for $n = 1$, $\cos(\text{arc } \cos(x)) = x$. Consequently,

- $T_0(x) = 1$,
- $T_1(x) = x$,
- $T_2(x) = 2x^2 - 1$ (note that $\cos(2x) = \cos^2 x - \sin^2 x = 2\cos^2 x - 1$),
- $T_3(x) = 4x^3 - 3x$,
- $T_4(x) = 8x^4 - 8x^2 + 1$,
- $\cdots$

A CP can be found using the previous two polynomials using the following recursive formula:

$$T_{n+1}(x) = 2xT_n(x) - T_{n-1}(x), \quad n = 1, 2, 3, \ldots. \tag{3.24}$$

Also, x can be expressed in terms of CP:

$$x^0 = T_0, \quad x^1 = T_1, \quad x^2 = \frac{1}{2}(T_0 + T_2), \quad x^3 = \frac{1}{4}(3T_1 + T_3).$$

Let us discuss using CP for approximating a function. In order to take advantage of CP, the sample points need to be selected between -1 and 1 as

$$x_k = \cos\frac{2N - 2k + 1}{2(N+1)}\pi \quad \text{for } k = 0, 2, \ldots, N. \tag{3.25}$$

For example, a domain is divided into three segments, $N = 3$. Hence, $x_k = \cos\frac{7-2k}{8}\pi$, $x_0 = -0.923879$, $x_1 = -0.382683$, $x_2 = 0.382683$, and $x_3 = 0.923879$. The CPs of degree up to N are orthogonal in a discrete sense on the interval $[-1, +1]$:

$$\sum_{k=1}^{N+1} T_i(x_k)T_j(x_k) = 0 \quad \text{for } i \neq j \text{ and } i \leqslant N, \qquad (3.26)$$

$$\sum_{k=1}^{N+1} T_i(x_k)T_j(x_k) = N+1 \quad \text{for } i = j = 0, \qquad (3.27)$$

$$\sum_{k=1}^{N+1} T_i(x_k)T_j(x_k) = \frac{1}{2}(N+1) \quad \text{for } i = j \quad \text{and} \quad 0 < i \leqslant N. \ (3.28)$$

The above orthogonality property of CP is very useful in calculating the interpolation coefficients.

A function, $f(x)$, can be approximated as $p_N(x)$ in term of CP as

$$p_N(x) = \sum_{i=0}^{N} c_i T_i(x). \qquad (3.29)$$

The coefficient c_i can be formulated as follows:

For $i = 0$,

$$c_0 = \frac{1}{N+1}\sum_{k=0}^{N} f(x_k)T_i(x_k) = \frac{1}{N+1}\sum_{k=0}^{N} f(x_k); \qquad (3.30)$$

for $i = 1$,

$$c_1 = \frac{2}{N+1}\sum_{k=0}^{N} f(x_k)T_i(x_k) = \frac{2}{N+1}\sum_{k=0}^{N} f(x_k)x_k; \qquad (3.31)$$

for $i = 2$,

$$c_1 = \frac{2}{N+1}\sum_{k=0}^{N} f(x_k)T_i(x_k) = \frac{2}{N+1}\sum_{k=0}^{N} f(x_k)(2x_k^2 - 1); \quad (3.32)$$

and so on.

Example: Approximate $f(z) = e^{-2z}$, in the range of $0 \leqslant z \leqslant 2$, using a second-order $(N = 2)$ CP. Compare the errors associated with Chebyshev and Lagrange interpolations.

The independent variable needs to be linearly transformed from a range of $[0, 2]$ to $[-1, 1]$, using formula (3.22), where $a = 0$ and $b = 2$. Hence,

$$z = x + 1. \tag{3.33}$$

So, function $f(z)$ can be written as $f(x) = e^{-2(x+1)}$.

Evaluate x_k:

$$x_k = \cos\left(\frac{2N - 2k + 1}{2(N + 1)}\right)\pi \quad \text{for } k = 0, 2, \ldots, N. \tag{3.34}$$

For $N = 2$, $x_0 = -0.866025$, $x_1 = 0.0$, and $x_2 = 0.866025$. The polynomial can be written as

$$P_2(x) = \sum_0^N c_k T(x_k). \tag{3.35}$$

The coefficient c_k evaluates as

$$c_0 = \frac{1}{3}\left[e^{-2(-0.866025+1)} + e^{-2(0+1)} + e^{-2(0.866025+1)}\right], \tag{3.36}$$

i.e.,

$$c_0 = \frac{1}{3}[0.7649460 + 0.1353353 + 0.0239437] = 0.308075 \tag{3.37}$$

and

$$c_1 = \frac{2}{3}[0.7649460 \times (-0.866025) + 0.1353353 \times 0.0$$
$$+ 0.0239437 \times (0.866025)]$$
$$= -0.427818. \tag{3.38}$$

To evaluate c_2, use the fact that $T_2(x_k) = 2x_k^2 - 1$. Hence,

$$T_2(x_0) = 2 \times (-0.866025)^2 - 1 = 0.5, \quad T_2(x_1) = 2 \times (0)^2 - 1 = -1,$$
and $\quad T_2(x_2) = 0.5.$

Therefore,

$$c_2 = \frac{2}{3}[0.7649460 \times 0.5 + 0.1353353 \times (-1) + 0.0239437 \times 0.5]$$

$$= 0.17274. \tag{3.39}$$

The polynomial is

$$P_2(x) = 0.308075 - 0.427818x + 0.17274(2x^2 - 1) \tag{3.40}$$

or

$$P_2(x) = 0.135335 - 0.427818x + 0.34548x^2. \tag{3.41}$$

In term of z,

$$P_2(z) = 0.308075 - 0.427818(z - 1) + 0.17274(2z^2 - 4z + 1), \tag{3.42}$$

which on rearranging reads

$$P_2(z) = 0.908633 - 1.11878z + 0.34548z^2. \tag{3.43}$$

Important note: If we use the same data of Chebyshev points ($x_0 = -0.866025$, $x_1 = 0$, $x_2 = 0.866025$) for Lagrange polynomial instead of a uniform interval, we end up with the same polynomial (CP). Let us check that,

for $x = -0.866025$, $e^{-2(x+1)} = 0.764946$,
for $x = 0$, $e^{-2(x+1)} = 0.135335$, and
for $x = 0.866025$, $e^{-2(x+1)} = 0.023944$. Hence,

$$P_l = \frac{(x - 0)(x - 0.866025)}{(-0.866025 - 0)(-0.866025 - 0.866025)}0.764946$$

$$+ \frac{(x + 0.866025)(x - 0.86625)}{(0 + 0.866025)(0 - 0.86625)}0.135335$$

$$+ \frac{(x + 0.866025)(x - 0)}{(0.866025 + 0.866025)(0.866025 - 0)}0.023944.$$

Hence,

$$P_l = x(x - 0.866025) \times 0.51 - (x + 0.866025)(x - 0.866025)$$

$$\times 0.1804 + x(x + 0.866025) \times 0.016. \tag{3.44}$$

Further simplification of the above equation yields

$$P_l = 0.1353 - 0.4278x + 0.3456x^2, \qquad (3.45)$$

which is exactly as $P_2(x)$ given in Eq. (3.41). In terms of z,

$$P_l = 0.9087 - 1.118z + 0.3456z^2. \qquad (3.46)$$

Figure 3.1 compares Chebyshev $(P_2(z))$ with Lagrange polynomial with uniform intervals $(z = 0.5, 1.0, 2.0)$ for the function $f = e^{-2z}$. It is clear that CP performs better than Lagrange polynomial, especially at the ends of the domain.

Regardless of the polynomial type, all of them yield the same results for the same number of points (and their location) and for the same order of the polynomial.

As a conclusion, using the interpolation of non-uniformly distributed points, preferably Chebyshev, we end up with the same polynomial, regardless of the process followed to get the polynomial we use. Also, using non-uniformly distributed points is preferable to minimize the error, especially at the end regions.

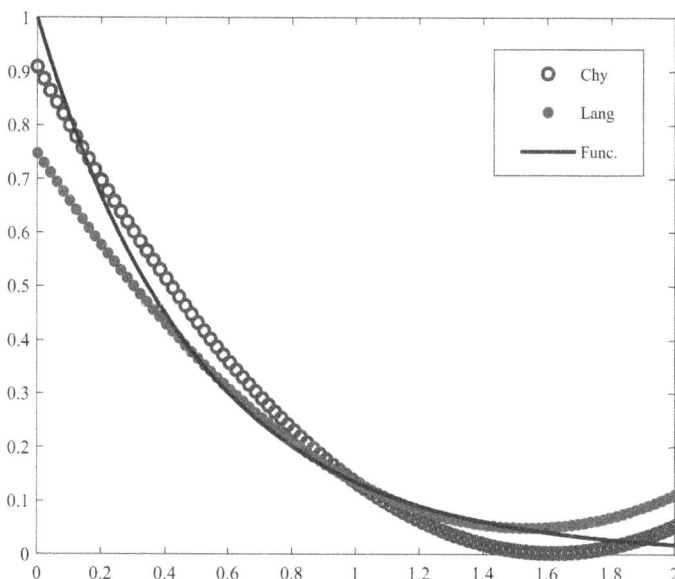

Figure 3.1: Comparison of the interpolated function with Lagrange and Chebyshev polynomials.

3.6　Least Square Method

3.6.1　One-variable fitting functions

A set of data $[x_i, y_i]$ is given and we are seeking a best fit to the data. The issue that may arise with any dataset is the lack of exactness due to error associated with the experimental measurements and numerical approximations. For want of anything better, the best fitting formula should minimize the error. For example, let us try to fit linearly a dataset:

$$y = a + bx. \tag{3.47}$$

The idea is to find the best values of a and b that minimize the error of

$$y_i = a + bx_i, \quad i = 1, \ldots, N, \tag{3.48}$$

where x_i and y_i, $i = 1, \ldots, N$, make up the dataset; a and b are constants to be determined. The error, e_i, is the difference between the measured values and the fitted values:

$$e_i = y_i - (a + bx_i). \tag{3.49}$$

The total errors squared (error vector length squared) is defined by

$$S(a, b) = \sum_{i=1}^{N}(e_i^2) = \sum_{i=1}^{N}(y_i - a - bx_i)^2. \tag{3.50}$$

To minimize the total error, the above equation needs to be differentiated partially with respect to a and b and set to zero. (This is the classical way to finding minima or maxima of a function.) The results of the differentiations and setting them to zero read

$$\frac{\partial S}{\partial a} = \sum_{i=1}^{N} -2(y_i - a - bx_i) = 0 \tag{3.51}$$

and

$$\frac{\partial S}{\partial b} = \sum_{i=1}^{N} -2x_i(y_i - a - bx_i) = 0. \tag{3.52}$$

Rearranging,

$$aN + b \sum_{i=1}^{N} x_i = \sum_{i=1}^{N} y_i, \tag{3.53}$$

$$a \sum_{i=1}^{N} x_i + b \sum_{i=1}^{N} x_i^2 = \sum_{i=1}^{N} y_i x_i. \tag{3.54}$$

Solving for a and b,

$$a = \frac{\sum_{i=1}^{N} y_i}{N} - b \frac{\sum_{i=1}^{N} x_i}{N} \tag{3.55}$$

and

$$b = \frac{N \sum_{i=1}^{N} x_i y_i - \left(\sum_{i=1}^{N} x_i \right) \left(\sum_{i=1}^{N} y_i \right)}{N \sum_{i=1}^{N} x_i^2 - \left(\sum_{i=1}^{N} x_i \right)^2}. \tag{3.56}$$

Like $y = a + bx$, different forms of functions may be reformulated. For example,

$$y = c_1 + c_2 e^x. \tag{3.57}$$

We define $X = e^x$, hence the equation formula will be $y = c_1 + c_2 X$.

Also, the formula, namely $y = ax^b$, can be written as $Y = A + bX$, where $Y = \ln(y)$, $A = \ln(a)$, and $X = \ln(x)$.

Formula $y = ae^{bx}$ can be written as $Y = A + bx$, where $Y = \ln(y)$, $A = \ln(a)$, and so on. If it is difficult to formulate a function as a linear function, we need to follow the above procedure to find the optimal constants.

3.6.2 Multi-variable fitting functions

One way to use a least square fitting for a multi-variable will be illustrated in the following example:

$$z = ax^n y^m. \tag{3.58}$$

Such a formula is very common in engineering, such as fluid mechanics and heat transfer, for example in correlating Nusselt number with

Reynolds and Prandtl numbers. Say for each given y, a correlation is developed as $z = cx^n$. Note that c is a function of y. The equation can be formulated as $\ln z = \ln c + n \ln x$. A set of such correlations is developed for each y, where c is function of y. Therefore, c is correlated with y following a similar procedure.

Another approach to deal with the above equation is by linearizing it as

$$\ln z = \ln a + n \ln x + m \ln y. \tag{3.59}$$

Hence, the equation can be written as

$$Z = A + nX + mY. \tag{3.60}$$

Therefore, the error is given by

$$S(A, n, m) = \sum_{i=1}^{N} e_i = \sum_{i=1}^{N} [Z_i - (A + nX_i + mY_i)]^2. \tag{3.61}$$

Differentiating partially with respect to each unknown (A, n, or m) and setting the differential to zero reads

$$\frac{\partial S}{\partial A} = -2 \sum_{i=1}^{N} [Z_i - (A + nX_i + mY_i)] = 0, \tag{3.62}$$

$$\frac{\partial S}{\partial n} = -2 \sum_{i=1}^{N} X_i [Z_i - (A + nX_i + mY_i)] = 0, \tag{3.63}$$

and

$$\frac{\partial S}{\partial m} = -2 \sum_{i=1}^{N} Y_i [Z_i - (A + nX_i + mY_i)] = 0. \tag{3.64}$$

The above equations can be formulated in the matrix form:

$$\begin{bmatrix} N & \sum X_i & \sum Y_i \\ \sum X_i & \sum X_i^2 & \sum X_i Y_i \\ \sum Y_i & \sum X_i Y_i & \sum Y_i^2 \end{bmatrix} \begin{bmatrix} A \\ n \\ m \end{bmatrix} = \begin{bmatrix} \sum Z_i \\ \sum X_i Z_i \\ \sum Y_i Z_i \end{bmatrix}.$$

The above set of linear equations can be solved by inverting the matrix or by using any of the methods discussed in Chapter 2.

3.7 Spline Interpolation

Using higher-order interpolation may lead to oscillations that inevitably increases the error associated with the interpolated values. One of the remedies is to split the domain into segments and apply lower-order polynomials for each segment. However, without special care, the segments do not smoothly connect at the interface points (knots). In other words, the polynomial is in general not differentiable at the interface points. To ensure smoothness at the interfaces, we need to impose the continuity of the derivatives at every point including the interface points (knots). This method is called spline method.

3.7.1 Quadratic splines

This section discusses the quadratic spline, which will pave the way toward explaining the more popular cubic splines, the procedure being the same.

3.7.1.1 *Basic quadratic splines*

Consider a set of data points, n, (Fig. 3.2 shows $n = 6$). Those data points lay between a and b. A quadratic polynomial,

$$p_k(x) = a_k + b_k x + c_k x^2, \qquad (3.65)$$

passes through two points (as shown in Fig. 3.2(a)) for each segment, k, which is always equal to $n - 1$, ($k = 5$, Fig. 3.2(a), for $n = 6$). Each segment has three unknowns (a_k, b_k, and c_k). Therefore, the number of unknowns is $3k$, i.e., $3(n-1)$. There are two points (knots) at each segment, hence the number of equations at each segment is 2, with the total of $2k$, i.e., $2(n - 1)$. The polynomial should have continuous first derivative at each interior points (knots), including the interfaces, which yields $k - 1$ equations.

$$p_i'(x_i) = p_{i+1}'(x_i), \quad i = 1, 2, \ldots, k - 1. \qquad (3.66)$$

Note that superscript ($'$) refers to differentiation. Now, let us summarize the number of unknowns and the number of equations:

Unknowns: $3(n - 1)$.
Number of equations: $2k + k - 1 = 3k - 1 = 3(n - 1) - 1$.

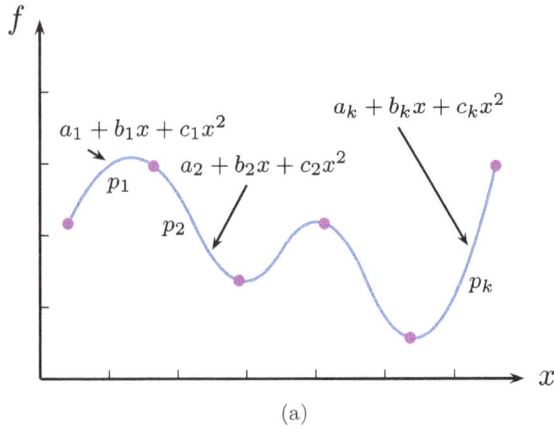

(a)

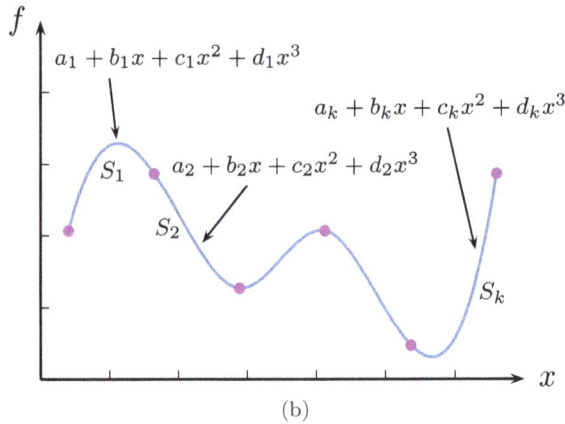

(b)

Figure 3.2:   Splines: (*a*) quadratic and (*b*) cubic.

To close the problem, an extra equation has to be considered. In most cases, the first or second derivative at the initial point ($x = a$) is set to zero.

Example: $(0, 1)$, $(1, 2.5)$, $(2, 3)$ are three points ($x = 0, 1, 2$) and the values of the function at those points are 1, 2.5, and 3, respectively. Therefore, $n = 3$ and $k = 2$. The number of unknowns is 6.

$$p_1 = a_1 + b_1 x + c_1 x^2 \qquad \text{on the interval } [0, 1]. \tag{3.67}$$

$$p_2 = a_2 + b_2 x + c_2 x^2 \qquad \text{on the interval } [1, 2]. \tag{3.68}$$

$$p_1(0) = 1 = a_1; \qquad\qquad p_1(1) = 2.5 = a_1 + b_1 + c_1; \quad (3.69)$$

$$p_2(1) = 2.5 = a_2 + b_2 + c_2; \qquad p_2(2) = 3 = a_2 + 2b_2 + 4c_2. \quad (3.70)$$

Above are four equations, $p_1(0), p_1(1), p_2(1)$, and $p_2(2)$. The continuity of the first derivative at the interface between p_1 and p_2 at $x = 1$ reads

$$p_1'(1) = b_1 + 2c_1 = p_2'(1) = b_2 + 2c_2. \qquad (3.71)$$

Let us impose the first derivative at $x = 0$ to zero:

$$p_1'(0) = b_1 = 0. \qquad (3.72)$$

Hence,

$$p_1 = 1 + 1.5x^2 \qquad (3.73)$$

and

$$p_2 = -3 + 8x - 2.5x^2. \qquad (3.74)$$

3.7.1.2 *Generalized quadratic splines*

The first derivative of a quadratic spline is a linear spline. Let

$$z_i = S'(x_i), \quad 0 \leqslant i \leqslant n. \qquad (3.75)$$

Therefore, $S'(x)$ is a straight line joining the points (x_{i-1}, z_{i-1}) and (x_i, z_i):

$$p_i'(x) = z_{i-1} + \frac{z_i - z_{i-1}}{h_i}(x - x_{i-1}), \qquad (3.76)$$

where $h_i = x_i - x_{i-1}$. Integrating the above equation yields

$$p_i(x) = z_{i-1}(x_i - x_{i-1}) + \frac{z_i - z_{i-1}}{2h_i}(x_i - x_{i-1})^2 + c_i, \qquad (3.77)$$

where c_i is a constant of integration. Since $p_i(x_i) = f(x_i)$ and $p_i(x_{i-1}) = f(x_{i-1})$, then

$$c_i = f(x_{i-1}), \quad i = 1, 2, 3, \ldots, n \qquad (3.78)$$

and

$$z_{i-1}h_i + \frac{z_i - z_{i-1}}{2}h_i + c_i = f(x_i), \quad i = 1, 2, 3, \ldots, n. \qquad (3.79)$$

Since $c_i = f(x_{i-1})$, then

$$\frac{z_i + z_{i-1}}{2} h_i = f(x_i) - f(x_{i-1}), \quad i = 1, 2, 3, \ldots, n. \tag{3.80}$$

The above equations on setting $p'_1(x_0) = 0$, i.e., $z_0 = 0$, yield a system of linear equations:

$$\mathbf{Az} = \mathbf{b}, \tag{3.81}$$

where $\mathbf{z} = [z_0, z_1, z_2, \ldots, z_n]^T$ and $\mathbf{b} = [b_0, b_1, b_2, \ldots, b_n]^T$.

$$b_i = \begin{cases} 0 & i = 0, \\ \dfrac{2(f(x_i) - f(x_{i-1}))}{x_i - x_{i-1}} & 1 \leqslant i \leqslant n, \end{cases} \tag{3.82}$$

and

$$\mathbf{A} = \begin{bmatrix} 1 & 0 & 0 & .. & .. & .. \\ 1 & 1 & 0 & 0 & .. & .. \\ 0 & 1 & 1 & 0 & .. & .. \\ 0 & 0 & 1 & 1 & 0 & .. \\ .. & .. & .. & .. & .. & .. \\ .. & .. & 0 & 0 & 1 & 1 \end{bmatrix}. \tag{3.83}$$

Solve for z_i. The splines can be calculated as,

$$p_i(x) = z_{i-1}(x_i - x_{i-1}) + \frac{z_i - z_{i-1}}{2h_i}(x_i - x_{i-1})^2 + f(x_{i-1}). \tag{3.84}$$

3.7.2   Cubic splines

A more popular method is the cubic spline method. In this method, a third-order polynomial needs to be passed through each segment (two points)

$$S_i(x) = a_i + b_i x + c_i x^2 + d_i x^3, \tag{3.85}$$

where a_i, b_i, c_i, and d_i are unknown coefficients which need to be determined. The first (slope) and second (curvature) derivatives of the polynomial are continuous everywhere in the domain including the knots.

Hence, a function $f(x)$ in the interval $[a, b]$ has known values of $f(x_i)$ at x_i. Therefore, for n points (knots), there are $n - 1$ segments. The number of unknowns is $4(n - 1)$. Each segment has two points, which leads to two equations as each segment with a total of $2(n - 1)$. The continuity of first and second derivatives at the interfaces leads to $2(n-2)$ equations. Hence, the total number of equations is $2(n - 1) + 2(n - 2) = 4(n - 1) - 2$. We are short of two equations to close the problem. In general, we can impose the first and/or second derivatives at $x = a$ and/or $x = b$ (or their combination) to zero, which is called natural cubic spline. Similar to the quadratic spline, we end up with a system of linear equations:

$$\mathbf{A}\mathbf{z} = \mathbf{b}, \tag{3.86}$$

where

$$\mathbf{A} = \begin{bmatrix} 2h_1 & h_1 & 0 & .. & .. & .. \\ h_1 & 2(h_1 + h_2) & h_2 & 0 & .. & .. \\ 0 & h_2 & 2(h_2 + h_3) & h_3 & .. & .. \\ 0 & 0 & h_3 & 2(h_3 + h_4) & h_4 & .. \\ .. & .. & .. & .. & .. & .. \\ .. & .. & 0 & 0 & h_n & 2h_n \end{bmatrix}, \tag{3.87}$$

$$b_i = \begin{cases} \dfrac{6}{h_1}(f_1 - f_0) - 6f_0' & i = 0, \\[2ex] 6f_n' - \dfrac{6}{h_n}(f_n - f_{n-1}) & i = n, \end{cases} \tag{3.88}$$

$$S_{i+1}(x) = \frac{z_i}{6h_{i+1}}(x_{i+1} - x)^3 + \frac{z_{i+1}}{6h_{i+1}}(x - x_i)^3$$
$$+ \left(\frac{f_{i+1}}{h_{i+1}} - \frac{z_{i+1}h_{i+1}}{6} \right)(x - x_i)$$
$$+ \left(\frac{f_i}{h_{i+1}} - \frac{z_i h_{i+1}}{6} \right)(x_{i+1} - x). \tag{3.89}$$

Note that matrix $\mathbf{A}$ is a tri-diagonal matrix. Therefore, $\mathbf{Az} = \mathbf{b}$ can be solved ideally using Thomas algorithm, see Section 2.4.2.

3.8   Problems

Problem 1: The constant pressure specific heat C_p and enthalpy h of low-pressure air are tabulated in Table 3.2. Using Lagrange polynomials with the base point as close to the specified value of T as possible, calculate

(a) $C_p(1120)$ using two points,
(b) $C_p(1120)$ using three points, and
(c) the best line that can approximate the data.

Problem 2: The pressure drop, ΔP, and the flow velocity, V, for incompressible flow in a pipe can be related as

$$\Delta P = -\tfrac{1}{2} f \rho V^2 (L/D), \tag{3.90}$$

where ρ, L, and D are fluid density, pipe length, and pipe diameter, respectively. f is the Darcy friction coefficient. For a laminar flow, the friction factor is related to the Reynolds number (Re) as

$$f = aRe^b. \tag{3.91}$$

Using the tabulated experimental data, as shown in Table 3.3, determine the best fit to calculate a and b.

Table 3.2:   Specific heat and enthalpy vs. temperature data.

T, K	C_p, kJ/kg K	h, kJ/kg	T, K	C_p, kJ/kg K	h, kJ/kg
1000	1.1410	1047.248	1400	1.1982	1515.792
1100	1.1573	1162.174	1500	1.2095	1636.188
1200	1.1722	1278.663	1600	1.2197	1757.657
1300	1.1858	1396.578			

Table 3.3:   Friction coefficient vs. Reynolds number data.

Re	500	1000	1500	2000
f	0.0320	0.0160	0.0107	0.0080

Problem 3: Explain how to apply the least square method to fit data with the following equations (for some cases, nonlinear systems are retrieved):

- $y(x) = \dfrac{a}{1 + bx}$.
- $y(x) = a + bx^n$ (n given).
- $y(x) = a + bx + cx^2$.
- $y(x) = a + be^x + ce^{2x}$.
- $y(x) = a \left(\ln x\right)^2 + b \ln x + c$.
- $y(x) = a \sin\left(2\pi f x\right) + b \cos\left(2\pi f x\right)$ (f given).
- $y(x) = a \sin\left(bx + c\right)$.
- $y(x) = a \exp\left(-\frac{1}{2}\left[\frac{x-b}{c}\right]^2\right)$.
- $z(x, y) = \displaystyle\sum_{k=0}^{d} x^k y^{d-k}$ (d given).

Chapter 4

Nonlinear Equations

In many engineering applications, solving nonlinear equations or finding the roots of equations is one of the most important computational problems, either directly or indirectly. Nonlinear equations arise in many engineering and science problems. Solving these equations is called finding their roots. In the optimization process, we need to find the critical point, for instance what is the x value (or what are the x values) at which the following curves intersect:

$$y_1 = x^2 \quad \text{and} \quad y_2 = x + 2 \tag{4.1}$$

or, equivalently, what are the values of x for which the ordinates y_1 and y_2 are equal:

$$x^2 = x + 2. \tag{4.2}$$

In order words, what are the roots of the the following function:

$$f(x) - x^2 - x - 2, \quad \text{where } f(x) = 0. \tag{4.3}$$

The general form of the above type of equations is

$$f(x) = ax^2 + bx + c. \tag{4.4}$$

For quadratic equations, such as Eq. (4.4), we can use the following analytical formula to find the roots:

$$x = \frac{-b \pm \sqrt{b^2 - 4ac}}{2a}. \tag{4.5}$$

However, there are no analytical formulae for most of the more complicated nonlinear equations. Therefore, we need to use numerical techniques to approximate the roots of equations. The reader may argue that instead of solving the above-mentioned equation, there are many plotting packages that can be used to plot the equation and visually figure out the roots. The argument is valid. However, it is not possible to extract accurate values visually. More importantly, for complicated problems with more than one variable, it is impossible to visualize or plot those equations. Furthermore, in more involved calculations, we may need the roots as part of the intermediate calculations. It is neither efficient nor desirable to stop a calculation (a computer program or subroutine) to find a root of an equation and feed that value to the calculation again. Therefore, there have to be seamless methods that calculate the root(s) of equation(s) when needed.

In the following sections, we discuss a few methods used to solve nonlinear equations.

4.1 Bisection Method

One basic method of finding a root of an equation is the bisection method. The root is confined in an interval between two points, say a and b, provided that the function is continuous and changes its sign. This is called the intermediate value theorem. The procedure is simple. Evaluate the value of the function at two points (a and b). If the value of the function changes sign, then at least one root lies between those points. Without loss of generality, let us say the value of $f(a)$ is negative and the value of $f(b)$ is positive. The procedure is to shorten the interval within which the root lies, see Fig. 4.1. This can be done by splitting the interval as

$$c = \frac{a+b}{2} \tag{4.6}$$

and evaluating the function at point c. If $f(c)$ is negative, the root is necessarily between c and b. Of course, if $f(c)$ is positive, the root is between a and c, and so on and so forth. The main drawback of the method is its slow convergence. However, it always converges and can be used as a starting point to other methods.

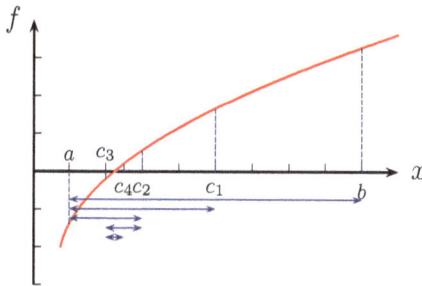

Figure 4.1: Illustration of the concept of bisection method.

The algorithm is as follows:

Evaluate the function at two points a and b, ensuring that the function changes its sign. Then,

- for $i = 0, 1, 2, 3, \ldots$
- $c = (a_i + b_i)/2$
- if $f(a_i)f(c) \leqslant 0$
    then $a_{i+1} = a_i$
         $b_{i+1} = c$
- else $a_{i+1} = c$
       $b_{i+1} = b_i$

4.2 Fixed Point Method

Another simple method is the fixed point method.

The equation to be solved is first cast as

$$x = f(x). \tag{4.7}$$

The initial guess, x_0, is substituted in the function f, for a better estimation of x:

$$x_1 = f(x_0) \tag{4.8}$$

and so on until convergence.

The convergence is guaranteed if the derivative of $f(x)$ with respect to x is less than one $\left(\left| \frac{df}{dx} \right| < 1 \right)$. Hence, first we need to identify the root location.

For example, find a root of the following equation:

$$x^2 - 3x + 2 = 0. \tag{4.9}$$

Let us test the sign of the left-hand side of this equation for $x = 0$ and $x = 1.5$.

For $x = 0$, $f(x) = x^2 - 3x + 2$ is $+2$; for $x = 1.5$, $f(x)$ is -0.25. Since the function changes its sign between 0 and 1.5, there is at least one root in the interval $[0; 1.5]$.

The equation can be cast in different ways as follows:

(a) $x = \frac{x^2+2}{3}$; the derivative of $\frac{x^2+2}{3}$ is $2x/3$, which is $2/3$ and 1 for $x = 0$ and 1.5, respectively;

or

(b) $x = \pm\sqrt{3x - 2}$; the derivative of $\sqrt{3x - 2}$ is $3/2(3x - 2)^{-0.5}$, which is not defined at $x = 0$. However, for $x = 1.5$, it is less than one;

or

(c) $x = \frac{-2}{x-3}$, and so on.

The question is which one of those converges? And if both of them converge, which one converges faster? Let us start with the initial guess of $x = 1.5$. Also, for b, consider the positive root.

a	b	c
1.41666	1.58114	1.33333
1.33564	1.65632	1.20000
1.26131	1.72730	1.11111
1.19697	1.78023	1.05880
1.14424	1.82775	1.03030
1.10310	1.86635	1.01538
1.07227	1.89712	1.00775

Notice that convergence of a is slow compared with convergence of c. Also, a and c converge to a root equal to 1, while b converges to a root equal to 2.

4.3 Newton–Raphson Method

Recall Taylor series:

$$f(x_{n+1}) = f(x_n) + \frac{df}{dx}(x_{n+1} - x_n) + \frac{1}{2}\frac{d^2 f}{dx^2}(x_{n+1} - x_n)^2 + O(\Delta x^3).$$

$$(4.10)$$

If $f(x_{n+1})$ is forced to be zero, then

$$0 = f(x_n) + \frac{df}{dx}(x_{n+1} - x_n) + \frac{1}{2}\frac{d^2 f}{dx^2}(x_{n+1} - x_n)^2 + O(\Delta x^3). \quad (4.11)$$

Dividing by $\frac{df}{dx}$, canceling the leading error $(-\frac{f''(x)}{f'(x)}(x_{n+1} - x_n)^2)$ and rearranging, the equation becomes

$$x_{n+1} = x_n - \frac{f(x_n)}{\left.\dfrac{df}{dx}\right|_{x_n}}. \qquad (4.12)$$

Hence, Newton–Raphson (N–R) method is nothing but Taylor series truncated at $O(\Delta x^2)$, which converges quadratically. Figure 4.2 illustrates N–R method graphically.

Notice that the last term on the right-hand side should be zero for a convergent solution. This recalls **SOR** method. Hence, in order to accelerate the convergence of N–R method, a relaxation factor ω

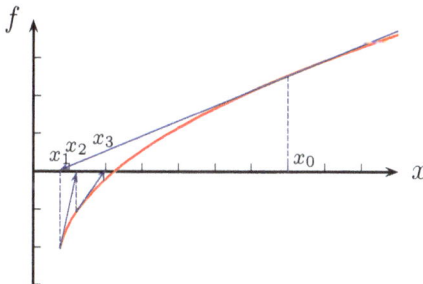

Figure 4.2: Illustration of the concept of Newton–Raphson method.

$(\omega > 1)$ can be considered, that is

$$x_{n+1} = x_n - \omega \frac{f(x_n)}{\left.\frac{df}{dx}\right|_{x_n}}. \tag{4.13}$$

It is also possible to use higher-order N–R method, e.g.,

$$f(x_{n+1}) = f(x_n) + \frac{df}{dx}(x_{n+1} - x_n)$$
$$+ \frac{d^2 f}{dx^2}\frac{(x_{n+1} - x_n)^2}{2!} + \cdots + O(\Delta x^3). \tag{4.14}$$

If $f(x_{n+1})$ is forced to be zero, then the following equation

$$0 = f(x_n) + \frac{df}{dx}(x_{n+1} - x_n) + \frac{d^2 f}{dx^2}\frac{(x_{n+1} - x_n)^2}{2!} \tag{4.15}$$

is solved for x_{n+1}. The problem is to solve a second-order nonlinear equation. The above quadratic equation, $ax^2 + bx + c = 0$, has the solution

$$x = \frac{-b \pm \sqrt{b^2 - 4ac}}{2a}. \tag{4.16}$$

Let us use the positive root and multiply and divide it by the conjugate:

$$\frac{-b + \sqrt{b^2 - 4ac}}{2a} = \frac{(-b + \sqrt{b^2 - 4ac}) * (b + \sqrt{b^2 - 4ac})}{2a(b^2 + \sqrt{b^2 - 4ac})}, \tag{4.17}$$

which yields

$$\frac{-2c}{b + \sqrt{b^2 - 4ac}}. \tag{4.18}$$

Therefore, Eq. (4.10) has a solution for x_{n+1} as

$$x_{n+1} = x_n - \frac{f(x_n)}{f'(x_n)\left[1 + \sqrt{1 - \frac{2f''(x_n)f(x_n)}{f'(x_n)^2}}\right]}. \tag{4.19}$$

Example: A spherical particle of diameter D is falling in oil. It experiences drag force, buoyancy force, and gravitational force.

When these forces are balanced, the particle reaches a constant velocity called the terminal velocity, which is equal to

$$u = \sqrt{\frac{4g(\rho_p - \rho_0)D}{3C_D\rho_0}}, \qquad (4.20)$$

where g, ρ_p, ρ_0, and C_D are gravity, density of the particle, density of oil, and the drag coefficient for creeping flow which is $C_D = \frac{24}{Re}$, where $Re = \frac{uD}{\nu}$ is Reynolds number and ν is kinematic viscosity of oil. Let $\nu = 0.001\,\mathrm{m^2/s}$, $\rho_0 = 850\,\mathrm{kg/m^3}$, $\rho_p = 2000\,\mathrm{kg/m^3}$, and $D = 0.002\,\mathrm{m}$. Hence, $Re = 2u$. and $C_D = 12/u$. Use N–R method to evaluate the thermal velocity.

Therefore, the terminal velocity satisfies the following equation:

$$u = 0.0543\sqrt{u}. \qquad (4.21)$$

If we put $f(u) = u - 0.0543u^{0.5}$, then $f' = 1 - 0.02715u^{-0.5}$ and N–R method yields

$$u_{i+1} = u_i - \frac{u - 0.0543u^{0.5}}{1 - 0.02715u^{-0.5}}. \qquad (4.22)$$

Let us set $u_0 = 0.01\,\mathrm{m/s}$. Then,

$$u_1 = 0.01 - \frac{0.1 - 0.0543\,(0.01)^{0.5}}{1 - 0.02715(0.01)^{-0.5}} = 0.003727 \qquad (4.23)$$

and $u_2 = 0.002985$, ... The exact value is 0.002948.

4.4   Chebyshev Method

Consider N–R method:

$$x_{n+1} = x_n - \frac{f(x_n)}{f'(x_n)}. \qquad (4.24)$$

Expand $f(x_{n+1})$ using Taylor series and force the function to be zero:

$$f(x_{n+1}) = f(x_n) + f'(x_n)(x_{n+1} - x_n) + \frac{1}{2}f''(x_n)(x_{n+1} - x_n)^2 = 0. \qquad (4.25)$$

From N–R equation, $x_{n+1} - x_n = -\frac{f(x_n)}{f'(x_n)}$, substitute the last term in the above equation:

$$f(x_n) + (x_{n+1} - x_n)f'(x_n) + \frac{1}{2}\frac{f(x_n)^2 f''(x_n)}{[f(x_n)']^2} = 0. \qquad (4.26)$$

On rearranging,

$$x_{n+1} = x_n - \frac{f(x_n)}{f'(x_n)} - \frac{1}{2}\frac{f(x_n)^2 f''(x_n)}{[f'(x_n)]^3}. \qquad (4.27)$$

The last term is correction of N–R equation. One of the advantages of this method is that the rate of convergence is cubic. However, we need to evaluate the second derivative.

4.5   Secant Method

The above-mentioned methods require the derivative of the function. In some applications, we do not have the explicit form of the function. For example, in experiments, the data are given at some points, and the function that relates those data is not known. In such cases, we use finite difference to evaluate the derivatives numerically. For example,

$$f'(x_n) \approx \frac{f(x_n) - f(x_{n-1})}{x_n - x_{n-1}}, \qquad (4.28)$$

in which case, Eq. (4.12) becomes

$$x_{n+1} = x_n - \frac{f(x_n)}{\dfrac{f(x_n) - f(x_{n-1})}{x_n - x_{n-1}}}. \qquad (4.29)$$

Therefore, we initially need two guess values for the function at two different points. This method is called **secant method**. It can be shown that if the method converges, it does so linearly.

4.6   Problems

Problem 1: Find the solutions of $e^{2.5x} - x = 7$.

Problem 2: The Beattie–Bridgeman gas equation of state is

$$P = \frac{RT}{V} + \frac{\alpha}{V^2} + \frac{\beta}{V^3} + \frac{\gamma}{V^4}, \tag{4.30}$$

where P, R, T, and V are the pressure, gas constant, absolute temperature, and volume, respectively. The parameters are

$$\alpha = bRT - a - \frac{cR}{T^2}, \tag{4.31}$$

$$\beta = -dbRT + ae - \frac{cbR}{T^2}, \tag{4.32}$$

$$\gamma = \frac{bdc}{T^2}, \tag{4.33}$$

where a, b, c, d, and e are constants that depend on the type of gas. To find a volume for a given pressure and temperature, N–R method can be used as follows:

$$f = \frac{RT}{V} + \frac{\alpha}{V^2} + \frac{\beta}{V^3} + \frac{\gamma}{V^4} - P, \tag{4.34}$$

$$f' = \frac{\partial f}{\partial V} = -\frac{RT}{V^2} - \frac{2\alpha}{V^3} - \frac{3\beta}{V^4} + \frac{4\gamma}{V^5}, \tag{4.35}$$

$$V_{k+1} = V_k - \frac{f}{f'}. \tag{4.36}$$

Problem 3: A beam is supported by its ends $x = 0$ and $x = L$. The beam is linearly loaded, as shown in Fig. 4.3. The beam deflection can be expressed as

$$y = \frac{FL}{120EI}(x^5 - 3L^2 x^3 + 2L^3 x^2),$$

where E and I are Young's modulus of elasticity and area moment of inertia, respectively. The above equation can be written in dimensionless form as

$$\eta = (\xi^5 - 3\xi^3 + 2\xi^2),$$

where $\eta = \frac{120E\,Iy}{F_L L^5}$ and $\xi = x/L$. Determine the location of maximum deflection.

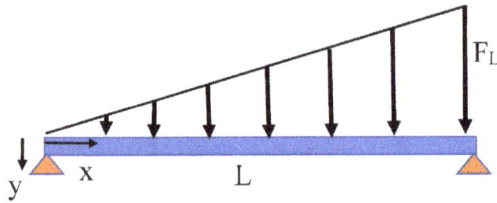

Figure 4.3:   A beam supported at its ends and linearly loaded.

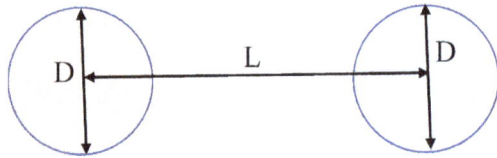

Figure 4.4:   Radiation between two infinite cylinders.

Problem 4: The radiation exchange between two surfaces can be written as

$$Q_{1,2} = \epsilon A F_{12}(T_1^4 - T_2^4),$$

where ϵ, A, and F_{12} are Stefan–Boltzmann constant, projected area, and shape factor, respectively.

For two identical, infinitely long cylinders (as shown in Fig. 4.4), the shape factor can be expressed as

$$F_{12} = \frac{1}{\pi}(\sin^{-1}(D/L) + \sqrt{(L/D)^2 - 1} - L/D).$$

Determine the ratio of L/D that yields $F_{12} = 0.5$.

Problem 5: The equation governing the kinematics of a rocket climbing vertically can be written as

$$v(t) = u \ln\left(\frac{m_0}{m_0 - \mu t}\right) - g\,t,$$

where $v(t)$ is the rocket speed, u is the relative speed of gases through the exhaust nozzle, m_0 is the initial rocket mass, μ is the mass flow rate of gases (due to fuel combustion), g is the acceleration of gravity (assumed constant), and t is the time. We set $u = 2500$ m/s, $m_0 = 2000$ kg, $\mu = 10$ kg/s and $g = 9.81$ m/s^2.

- Determine (if it exists) the time necessary for the rocket to reach the speed of sound at standard conditions, namely 340 m/s.
- We assume that the initial mass of the fuel is half the total mass of the rocket. Can the rocket reach the speed of 1000 m/s?

4.7 System of Nonlinear Algebraic Equations

In this section, we discuss solving a system of nonlinear equations using N–R method.

Taylor series is used to linearize functions with more than one variable. Consider three equations, $f(x_1, x_2, x_3)$, $g(x_1, x_2, x_3)$, and $h(x_1, x_2, x_3)$. On using Taylor series expansion up to the first terms,

$$f(x_1, x_2, x_3)^{k+1} = f(x_1, x_2, x_3)^k + \left.\frac{\partial f}{\partial x_1}\right|_k (x_1^{k+1} - x_1^k)$$

$$+ \left.\frac{\partial f}{\partial x_2}\right|_k (x_2^{k+1} - x_2^k) + \left.\frac{\partial f}{\partial x_3}\right|_k (x_3^{k+1} - x_3^k), \quad (4.37)$$

$$g(x_1, x_2, x_3)^{k+1} = g(x_1, x_2, x_3)^k + \left.\frac{\partial g}{\partial x_1}\right|_k (x_1^{k+1} - x_1^k)$$

$$+ \left.\frac{\partial g}{\partial x_2}\right|_k (x_2^{k+1} - x_2^k) + \left.\frac{\partial g}{\partial x_3}\right|_k (x_3^{k+1} - x_3^k), \quad (4.38)$$

$$h(x_1, x_2, x_3)^{k+1} = h(x_1, x_2, x_3)^k + \left.\frac{\partial h}{\partial x_1}\right|_k (x_1^{k+1} - x_1^k)$$

$$+ \left.\frac{\partial h}{\partial x_2}\right|_k (x_2^{k+1} - x_2^k) + \left.\frac{\partial h}{\partial x_3}\right|_k (x_3^{k+1} - x_3^k), \quad (4.39)$$

where the superscript k refers to the iteration number. Setting the left-hand side to zero and rearranging those equations yields

$$0 = f(x_1, x_2, x_3)^k + \left.\frac{\partial f}{\partial x_1}\right|_k \Delta x_1 + \left.\frac{\partial f}{\partial x_2}\right|_k \Delta x_2 + \left.\frac{\partial f}{\partial x_3}\right|_k \Delta x_3,$$

$$(4.40)$$

$$0 = g(x_1, x_2, x_3)^k + \left.\frac{\partial g}{\partial x_1}\right|_k \Delta x_1 + \left.\frac{\partial g}{\partial x_2}\right|_k \Delta x_2 + \left.\frac{\partial g}{\partial x_3}\right|_k \Delta x_3,$$

$$(4.41)$$

$$0 = h(x_1, x_2, x_3)^k + \left.\frac{\partial h}{\partial x_1}\right|_k \Delta x_1 + \left.\frac{\partial h}{\partial x_2}\right|_k \Delta x_2 + \left.\frac{\partial h}{\partial x_3}\right|_k \Delta x_3.$$

(4.42)

Or in matrix form,

$$\begin{bmatrix} \frac{\partial f}{\partial x_1} & \frac{\partial f}{\partial x_2} & \frac{\partial f}{\partial x_3} \\ \frac{\partial g}{\partial x_1} & \frac{\partial g}{\partial x_2} & \frac{\partial g}{\partial x_3} \\ \frac{\partial h}{\partial x_1} & \frac{\partial h}{\partial x_2} & \frac{\partial h}{\partial x_3} \end{bmatrix} \begin{bmatrix} \Delta x_1^k \\ \Delta x_2^k \\ \Delta x_3^k \end{bmatrix} = - \begin{bmatrix} f^k \\ g^k \\ h^k \end{bmatrix}.$$

where $\Delta x_i^k = (x_i^{k+1} - x_i^k)$, $i = 1, 2, 3$, which in turn can be cast in a compact form as $\mathbf{J}\Delta\mathbf{x}^k = \mathbf{f}^k$. Here, $\mathbf{f}^k$ stands for $(f^k, g^k, h^k)^T$ and $\mathbf{J}$ is called the Jacobian (or Jacobi) matrix. Since, for a given iteration, the above equations are linear, we can use any of the previously mentioned methods of solving the linear algebraic equations and solve the problem iteratively.

Example:

Solve the following system of nonlinear equations by N–R method:

$$e^{2x_1} - x_2 = 4,$$

(4.43)

$$x_2 - x_3^2 = 1,$$

(4.44)

$$x_3 - \sin x_1 = 0.$$

(4.45)

Solution: Let $f_1 = e^{2x_1} - x_2 - 4$, $f_2 = x_2 - x_3^2 - 1$, and $f_3 = x_3 - \sin x_1$.

Derivatives:

$f_{1,x_1} = 2e^{2x_1}$, $f_{1,x_2} = -1$, and $f_{1,x_3} = 0$.
$f_{2,x_1} = 0$, $f_{2,x_2} = 1$, and $f_{2,x_3} = 2x_3$.
$f_{3,x_1} = -\cos x_1$, $f_{3,x_2} = 0$, and $f_{3,x_3} = 1$.

Hence,

$$\begin{bmatrix} 2e^{2x_1} & -1 & 0 \\ 0 & 1 & 2x_3 \\ -\cos x_1 & 0 & 1 \end{bmatrix} \begin{bmatrix} \Delta x_1^k \\ \Delta x_2^k \\ \Delta x_3^k \end{bmatrix} = - \begin{bmatrix} f_1^k \\ f_2^k \\ f_3^k \end{bmatrix}.$$

Let us guess $x_1^0 = x_2^0 = x_3^0 = 1$. Therefore, subscript $k = 0$. We need to evaluate the functions,

$$f_1^0 = e^2 - 1 + 4 = 10.389, \quad f_2^0 = -1, \quad \text{and} \quad f_3^0 = 0.15853.$$

Hence,

$$
\begin{bmatrix} 2e^2 & -1 & 0 \\ 0 & 1 & 2 \\ -\cos 1 & 0 & 1 \end{bmatrix} \begin{bmatrix} \Delta x_1^0 \\ \Delta x_2^0 \\ \Delta x_3^0 \end{bmatrix} = - \begin{bmatrix} 10.389 \\ -1 \\ 0.15853 \end{bmatrix}.
$$

Solve for Δx_1^0, Δx_2^0, and Δx_3^0 using Gauss elimination process: $\Delta x_1^0 = 0.572$, $\Delta x_2^0 = -1.935$, and $\Delta x_3^0 = 0.4676$.

Recall that $\Delta x_1^0 = x_1^1 - x_1^0 = x_1^1 - 1$, hence the updated value of $x_1^1 = 1.572$. Similarly, $x_2^1 = -0.935$ and $x_3^1 = 1.4676$. These updated values will be used in evaluating the functions and their derivatives. The procedure is repeated until convergence is reached, that is,

$$
\sqrt{\frac{1}{N} \sum_{i=1}^{N} \left(\Delta x_i^k \right)^2} < \epsilon, \tag{4.46}
$$

where N is the number of equations and ϵ is the tolerance (a small number set by the user).

4.8   Problems

Problem 1: The pressure and temperature ratios of detonation wave moving toward a fresh gaseous fuel–air mixture can be expressed as

$$
\gamma m_r P_r^2 / T_r - (\gamma + 1) P_r + 1 = 0
$$

and

$$
\frac{\Delta H}{c_p T_1} + T_r - 1 = \frac{(\gamma - 1) m_r}{2\gamma} (P_r - 1)(1 + \frac{T_r}{m_r P_r}),
$$

where γ, m_r, ΔH, and T_1 are specific heat, molecular ratio, heat of reaction, and initial temperature (K), respectively. $P_r = P_2/P_1$ and $T_r = T_2/T_1$, where subscripts 1 and 2 refer to the initial and final values, respectively. For the following values, calculate P_r and T_r:

$$
\gamma = 1.3, \Delta H = -6000 \, \text{kcal/kg},
$$

$$
m_r = 18, T_1 = 300 \, \text{K}, \quad \text{and} \quad P_1 = 1.0 \, \text{atm}.
$$

Problem 2: During the conversion of methane to syngas with reduced oxygen, the following reactions take place:

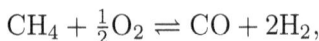

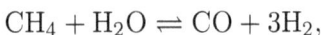

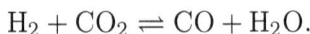

$$CH_4 + \tfrac{1}{2}O_2 \rightleftharpoons CO + 2H_2,$$

$$CH_4 + H_2O \rightleftharpoons CO + 3H_2,$$

$$H_2 + CO_2 \rightleftharpoons CO + H_2O.$$

The equilibrium constants for those reactions are

$$K_1 = \frac{P_{CO}P_{H_2}^2}{P_{CH_4}P_{O_2}^{0.5}} = 1.3 \times 10^{11},$$

$$K_2 = \frac{P_{CO}P_{H_2}^3}{P_{CH_4}P_{H_2O}} = 1.7837 \times 10^5,$$

$$K_3 = \frac{P_{CO}P_{H_2O}}{P_{CO_2}P_{H_2}} = 2.6058,$$

where P is the partial pressure.

Using energy balance for reactants combined with the above equations leads to a system of seven nonlinear algebraic equations:

$$\frac{1}{2}y_1 + y_2 + \frac{1}{2}y_3 - \frac{y_6}{y_7} = 0,$$

$$y_3 + y_4 + 2y_5 - \frac{2}{y_7} = 0,$$

$$y_1 + y_2 + y_5 - \frac{1}{y_7} = 0,$$

$$-28840y_1 - 139010y_2 - 78213y_3 + 18930y_4 + 8430y_5$$
$$+ \frac{13490}{y_7} - \frac{10690y_6}{y_7} = 0,$$

$$y_1 + y_2 + y_3 + y_4 + y_5 = 1,$$

$$400y_1y_4^3 - 1.7837x10^5 y_3y_5 = 0,$$

$$y_1y_3 - 2.8058y_2y_4 = 0,$$

where y stands for mole fraction of the species. The subscripts 1, 2, 3, 4, 5, 6, and 7 stand for mole fractions of $CO, CO_2, H_2O, H_2, CH_4, O_2$, and model number of moles of product gases per mole of CH_4 in the feed gases, respectively.

Solve those equations using N–R method. Use the following mole fractions as initial guesses: $y_1 = y_4 = y_6 = 0.5$, $y_2 = y_3 = y_5 = 0$, and $y_7 = 2.0$.

Problem 3: Lorenz problem, which describes atmospheric non-isothermal flow under equilibrium state, reads

$$x(\rho - z) - x = 0,$$
$$x^2 - \beta z = 0,$$

where x and z are the quantities linked with flow rate and vertical temperature gradient, respectively. We set $\rho = 20$ and $\beta = \frac{8}{3}$.

Starting with $(x^0, z^0) = (\beta, \rho)$, use N–R method to solve the nonlinear problem.

Chapter 5

Numerical Differentiation and Integration

5.1 Introduction to Numerical Differentiation

The results of most experimental and numerical analyses are discrete data. For instance, velocity, temperature, pressure, deflection, etc. are all measured at points or predicted at discrete locations in the solution domain. However, there may be a need to evaluate derivatives, for example, to find shear stress on a plate with a fluid flowing over it. Shear stress, τ, can be expressed as

$$\tau = \mu \frac{du}{dy}, \tag{5.1}$$

where μ, u, and y are viscosity, velocity, and normal distance to the plate, respectively. But to get τ, the derivative $\frac{du}{dy}$ needs to be evaluated at the plate surface. However, u as a function of y is not known. Only values of velocity are given at different and specific y locations. This is a typical example where numerical differentiation has a role to play. In heat transfer problems, the temperature can be easily measured at different locations. However, the heat flux, q'', defined by

$$q'' = -k \frac{dT}{dy}, \tag{5.2}$$

where k and T are thermal conductivity and temperature, respectively, is not. Moreover, the finite difference method, a method that converts differential equations (ordinary and partial) to algebraic equations, is intrinsically based on numerical differentiation. The backbone of numerical differentiation is Taylor series.

5.2 First Derivative

This section deals with the first derivative approximations. There are many ways to deal with the approximation.

5.2.1 Forward, backward, and central first-order approximations

Let us say that the values of the dependent function $f(x)$ are known at discrete points. Those points, x_i, x_{i+1}, x_{i+2}, ..., are equally spaced by Δx (see Fig. 5.1), at which $f(x)$ equals $f(x_i)$, $f(x_{i+1})$, $f(x_{i+2})$, ..., respectively. The function $f(x_{i+1})$ can be expanded at an arbitrary point x_i as far as we need to evaluate the derivative at x_i:

$$f(x_{i+1}) = f(x_i) + \left.\frac{df}{dx}\right|_{x_i} \Delta x + \frac{1}{2!}\left.\frac{d^2 f}{dx^2}\right|_{x_i} \Delta x^2 + \frac{1}{3!}\left.\frac{d^3 f}{dx^3}\right|_{x_i} \Delta x^3 + \cdots$$

$$+ \frac{1}{n!}\left.\frac{d^n f}{dx^n}\right|_{x_i} \Delta x^n. \tag{5.3}$$

The above expansion can be truncated to $O(\Delta x^2)$ as

$$f(x_{i+1}) = f(x_i) + \left.\frac{df}{dx}\right|_{x_i} \Delta x + O(\Delta x^2). \tag{5.4}$$

The term $O(\Delta x^2)$ means that the leading error is of the order of magnitude of Δx^2. Therefore,

$$\left.\frac{df}{dx}\right|_{x_i} = \frac{f(x_{i+1}) - f(x_i)}{\Delta x} + O(\Delta x). \tag{5.5}$$

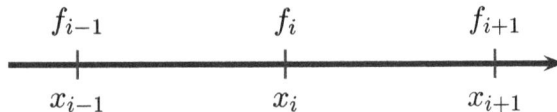

Figure 5.1: Equally spaced points.

In Eq. (5.5), the error order of magnitude is reduced from second order to first order because Eq. (5.4) is divided by Δx. The approximation is called **forward (FW) first-order approximation.**

It is also possible to expand the function backward:

$$f(x_{i-1}) = f(x_i) - \frac{df}{dx}\bigg|_{x_i} \Delta x + \frac{1}{2!}\frac{d^2 f}{dx^2}\bigg|_{x_i} \Delta x^2 - \frac{1}{3!}\frac{d^3 f}{dx^3}\bigg|_{x_i} \Delta x^3 + \cdots$$

$$+ \frac{1}{n!}\frac{d^n f}{dx^n}\bigg|_{x_i} \Delta x^n. \tag{5.6}$$

The above equation can be truncated to $O(\Delta x^2)$ as

$$f(x_{i-1}) = f(x_i) - \frac{df(x_i)}{dx}\Delta x + O(\Delta x^2). \tag{5.7}$$

Likewise, the term $O(\Delta x^2)$ means that the leading error is of the order of magnitude of Δx^2. Therefore,

$$\frac{df}{dx}\bigg|_{x_i} = \frac{f(x_i) - f(x_{i-1})}{\Delta x} + O(\Delta x). \tag{5.8}$$

The approximation is called **backward (BW) first-order approximation.** Note that both forward and backward approximations are first-order accurate. It is possible to increase the order of accuracy by subtracting Eq. (5.6) from Eq. (5.3):

$$f(x_{i+1}) - f(x_{i-1}) = 2\frac{df(x)}{dx}\big|_{x_i}\Delta x + \frac{2}{3!}\frac{d^3 f(x)}{dx^3}\big|_{x_i}\Delta x^3 + \cdots. \tag{5.9}$$

Rearranging the above equation yields

$$\frac{df}{dx}\bigg|_{x_i} = \frac{f(x_{i+1}) - f(x_{i-1})}{2\Delta x} + O(\Delta x^2). \tag{5.10}$$

The above equation is called **central difference (CD)**, which is second-order accurate. The different schemes used to approximate the first derivative of function f are illustrated in Fig. 5.2.

Example: Given $f(x) = x + 2x^2 - x^3$, approximate the first derivative of the function at points $x = 0$ and $x = 3$ using forward, backward, and central difference. Compare the approximate values with exact values for $\Delta x = 0.5$ and 0.1.

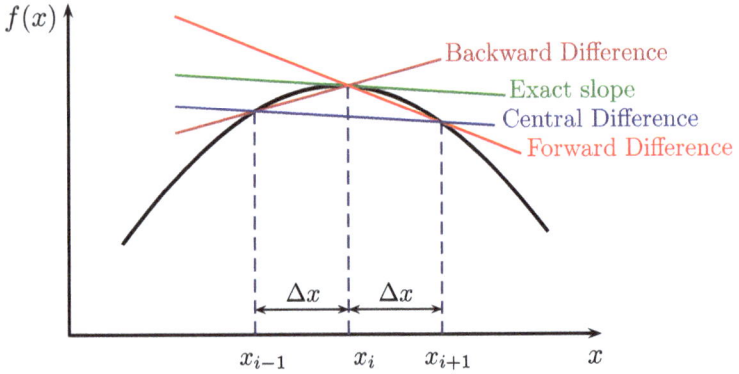

Figure 5.2:   Graphical interpretation of the various difference schemes.

Solution: The exact expression of $df/dx = 1 + 4x - 3x^2$. Hence, the exact value of the derivative at $x = 0$ and $x = 3$ are 1 and -14, respectively. For $\Delta x = 0.5$ and $x_i = 0$, the approximate values of the derivative using

FW is

$$\frac{df}{dx} = \frac{f(0+0.5) - f(0)}{\Delta x} = \frac{0.875 - 0}{0.5} = 1.75, \qquad (5.11)$$

BW is

$$\frac{df}{dx} = \frac{f(0) - f(0-0.5)}{\Delta x} = \frac{0 - (-0.625)}{0.5} = 1.25, \qquad (5.12)$$

CD is

$$\frac{df}{dx} = \frac{f(0+0.5) - f(0-0.5)}{2\Delta x} = \frac{0.875 - (-0.125)}{2 \times 0.5} = 1.0. \qquad (5.13)$$

For $\Delta x = 0.1$ and $x_i = 0$, the approximate values of the derivative using

FW is

$$\frac{df}{dx} = \frac{f(0+0.1) - f(0)}{\Delta x} = \frac{0.119 - 0}{0.1} = 1.19, \qquad (5.14)$$

BW is

$$\frac{df}{dx} = \frac{f(0) - f(0-0.1)}{\Delta x} = \frac{0 - (-0.081)}{0.1} = 0.81, \qquad (5.15)$$

CD is

$$\frac{df}{dx} = \frac{f(0+0.1) - f(0-0.1)}{2\Delta x} = \frac{0.119 - (-0.081)}{2 \times 0.1} = 1.0.$$
(5.16)

Let us evaluate the derivative at $x_i = 3$. For $\Delta x = 0.5$, using FW, BW, and CD yields -17.75, -13.25, and -14.25, respectively. For $\Delta x = 0.1$, using FW, BW, and CD yields -14.71, -13.31, and -14.01, respectively. It is clear that the CD yields better approximations because it is second-order accurate. In general, second-order accurate schemes yield better approximations.

Graphically, the FW, BW, and CD can be illustrated as in Fig. 5.2.

5.2.2 Higher-order approximations for first derivative

There are many ways to increase the order of accuracy by involving more points, either from the left side, from the right side of the given point, or from both. This can mainly be done by using Taylor series manipulations. For instance, higher-order FW schemes can include points only from the right side or more points from the right than the left side, and vice versa for BW schemes. The following is a list of the first derivative approximations including more points:

$$\left.\frac{df}{dx}\right|_i = f_x(i) = \frac{-3f_i + 4f_{i+1} - f_{i+2}}{2\Delta x} + O(\Delta x^2),$$
(5.17)

$$f_x(i) = \frac{f_{i-2} - 4f_{i-1} + 3f_i}{2\Delta x} + O(\Delta x^2),$$
(5.18)

$$f_x(i) = \frac{-2f_{i-1} - 3f_i + 6f_{i+1} - f_{i+2}}{6\Delta x} + O(\Delta x^3),$$
(5.19)

where $f_x(i)$ is a shorthand form of the derivative $\left.\frac{df}{dx}\right|_i$, and so on.

5.3 Higher-Order Derivatives

Similar to the first-order derivative, we can approximate second, third, fourth, and higher derivatives as we did for the approximation of the first order, i.e., by expanding Taylor series. For example, approximating the second derivative $(f_{xx}(i) = \left.\frac{d^2 f}{dx^2}\right|_i)$ can be done by adding side by side FW and BW Taylor expansions, Eqs. (5.6) and (5.3), which yields

$$f_{xx}(i) = \frac{f_{i+1} - 2f_i + f_{i-1}}{\Delta x^2} + O(\Delta x^2). \tag{5.20}$$

Including more points in Taylor expansion, we can approximate the second derivatives at point i by using points $(f_{i+1}, f_{i+2},$ and $f_{i+3})$, which yields

$$f_{xx} = \frac{2f_i - 5f_{i+1} + 4f_{i+2} - f_{i+3}}{\Delta x^2} + O(\Delta x^2). \tag{5.21}$$

Higher-order accurate schemes can be constructed by considering more points, for example,

$$f_{xx} = \frac{-f_{i-2} + 16f_{i-1} - 30f_i + 16f_{i+1} - f_{i+2}}{12\Delta x^2} + O(\Delta x^4), \tag{5.22}$$

and so on.

Notes:

- It is possible to check the consistency (or 0th-order accuracy) of the above equations by setting the function values to one, the results of which should be zero. For example, setting the values of the function to one in the last approximation yields $-1 + 16 - 30 + 16 - 1 = 0$.
- In approximating even-order derivatives (such as 2nd, 4th, and 6th-order derivatives), it is recommended to consider equal points from the main expansion point. For example, in approximating second derivative at point i, it is preferable to include points $i-1$ and $i+1$ and not points $i+1$ and $i+2$, and so on. Of course, mathematically, it does not make any difference. However, physically, even-order derivatives usually refer to diffusion processes which are essentially isotropic. The approximation should reflect such a feature by using neighbors equally weighted. In general, the odd derivatives are biased, i.e., directional: From an accuracy standpoint, no special care is needed. (But, as we will see in the chapters dealing with partial differential equations, we must pay much more attention from a stability point of view.)

5.4   Undetermined Coefficient Method

This method is very general and easy to apply. As discussed in the previous sections, in order to increase the accuracy of the approximation, more points need to be involved. The points can be selected either from the right or left side of the main point or from both sides. However, at the boundaries, some restrictions exist.

In general, we can express any order of derivative as

$$f(x)_n = \sum_{i=l}^{m} c_i f_i + O(), \tag{5.23}$$

where subscript n is the order of differentiation ($f(x)_0$ is by convention $f(x)$ itself), i is the point of interest, and l and m are starting and ending points, respectively. For example, if it is intended to find the first derivative of a function f at point i by including points from $i-1$ to $i+2$, then $n = 1$, $l = i-1$, and $m = i+2$. Therefore, four unknown coefficients, c_i, need to be determined. The procedure is to:

- expand the function f at x_l, x_{l+1}, ..., x_{m-1}, and x_m about the point x_i using Taylor series and to introduce these expansions into the ansartz (5.23).
- collect the coefficients of $f(x_i)_j$ ($j = 0, 1, \ldots$) and identify by comparing the left- and right-hand terms.

Example 1: Approximate $\frac{df}{dx}$ using the points x_i and x_{i+1}. Show the order of accuracy. Also, approximate the derivative using points x_i, x_{i-1}, and x_{i+1}. Show the order of accuracy. Assume that the points are uniformly spaced with space step h.

Solution:

$$f_{xi} = af_i + bf_{i+1}. \tag{5.24}$$

- Expand f at x_{i+1} about x_i using Taylor series, and substitute in the above equation:

$$f_{xi} = af_i + b(f_i + f_{xi}h + f_{xxi}h^2/2 + f_{xxxi}h^3/6$$
$$+ f_{xxxxi}h^4/24 + \cdots). \tag{5.25}$$

- The next step is to collect the coefficients of f_i, f_{xi}, f_{xxi}, f_{xxxi}, ... In other words, compare the coefficients of the left-hand side of the equation with the right-hand side of the equation. Hence, the above equation yields:

(1) Coefficient of f_i: $0 = a + b$ (note that there is no f_i on the left-hand side of the equation).
(2) Coefficient of f_x: $1 = bh$ (note that the coefficient of f_x on the left-hand side of the equation is 1 while on the right-hand side of the equation is bh).
(3) Coefficient of f_{xx}: $0 = bh^2/2$, which cannot be satisfied.
(4) Coefficient of f_{xxx}: $0 = bh^3/6$, which cannot be satisfied either.

Since we have only two coefficients a and b to determine, then coefficients of f_{xxi}, f_{xxxi}, ... may not be zero. Solving the first two equations yields $b = 1/h$ and $a = -1/h$. Therefore, Eq. (5.24) becomes

$$f_{xi} = \frac{f_{i+1} - f_i}{h}. \qquad (5.26)$$

The leading error is given by the term involved in the first unsatisfied equation, namely $bf_{xx}h^2/2$. Since $b = 1/h$, the error is $O(h)$ accordingly.

For part two of the example,

$$f_{xi} = af_i + bf_{i+1} + cf_{i-1}. \qquad (5.27)$$

Similarly, using Taylor expansion,

$$f_{xi} = af_i + b(f_i + f_{xi}h + f_{xxi}h^2/2 + f_{xxxi}h^3/6 + \cdots)$$
$$+ c(f_i - f_{xi}h + f_{xxi}h^2/2 - f_{xxxi}h^3/6 + \cdots). \qquad (5.28)$$

Equating term by term the coefficients of right-hand side with left-hand side:

(1) Coefficient of f_i: $0 = a + b + c$.
(2) Coefficient of f_{xi}: $1 = bh - ch$.
(3) Coefficient of f_{xxi}: $0 = bh^2/2 + ch^2/2$.
(4) Coefficient of f_{xxxi}: $0 = bh^3/6 - ch^3/6$.

Since we have three coefficients, a, b, and c, three equations need to be solved. The equation for coefficient of f_{xxi} balance leads to $c = -b$. Using other balance equations leads to $b = 1/2h$ and $a = 0$. Thus,

$$f_{xi} = \frac{f_{i+1} - f_{i-1}}{2h}. \qquad (5.29)$$

Since $b = -c$, there is no way to eliminate the coefficient of f_{xxx}. The coefficient of f_{xxx} is $bh^3/6$ and, since $b = 1/2h$, the leading error is $O(h^2)$ accordingly.

Example: It is intended to approximate the first derivative, $\frac{df}{dx}$, at point i by including f_{i-1}, f_i, f_{i+1}, and f_{i+2}. Hence, $n = 1$, $l = i - 1$, and $m = i + 2$. So,

$$\left(\frac{df}{dx}\right)_i = c_1 f_{i-1} + c_2 f_i + c_3 f_{i+1} + c_4 f_{i+2}. \qquad (5.30)$$

Note that c_1, c_2, c_3, and c_4 are used instead of c_{i-1}, c_i, c_{i+1}, and c_{i+2} for simplicity of notation. The second step is expanding f at x_{i-1}, x_{i+1}, and x_{i+1} about x_i using Taylor series and substituting in the above equation, which leads to

$$\begin{aligned} f_{xi} = {} & c_1(f_i - f_{xi}h + f_{xxi}h^2/2! - f_{xxxi}h^3/3! + \cdots) + c_2 f_i \\ & + c_3(f_i + f_{xi}h + f_{xxi}h^2/2! + f_{xxxi}h^3/3! + \cdots) \\ & + c_4(f_i + f_{xi}(2h) + f_{xxi}(2h)^2/2! + f_{xxxi}(2h)^3/3! + \cdots). \end{aligned}$$
$$(5.31)$$

The next step is to collect the coefficients of f_i, f_{xi}, f_{xxi}, f_{xxxi}, etc. Hence, the above equation yields

$$\begin{aligned} f_{xi} = {} & (c_1 + c_2 + c_3 + c_4)f_i + (-c_1 h + c_3 h + 2c_4 h)f_{xi} \\ & + (c_1 h^2/2! + c_3 h^2/2! + c_4(2h)^2/2!)f_{xxi} \\ & + (-c_1 h^3/3! + c_3 h^3/3! + c_4(2h)^3/3!)f_{xxxi} + \cdots. \end{aligned} \qquad (5.32)$$

In the above equation, the left-hand and the right-hand sides must equal term by term. So, the coefficient of f_i and of its derivatives must be equal too. Hence,

(1) Coefficient of f_i: $0 = c_1 + c_2 + c_3 + c_4$.
(2) Coefficient of f_{xi}: $1 = -c_1 h + c_3 h + 2c_4 h$.
(3) Coefficient of f_{xxi}: $0 = c_1 h^2/2 + c_3 h^2/2 + c_4 2h^2$.
(4) Coefficient of f_{xxxi}: $0 = -c_1 h^3/6 + c_3 h^3/6 + c_4 4h^3/3$.

The above equations can be cast in matrix form as

$$
\begin{bmatrix}
1.0 & 1.0 & 1.0 & 1.0 \\
-1.0 & 0.0 & 1.0 & 2.0 \\
1/2 & 0.0 & 1/2 & 2 \\
-1/6 & 0.0 & 1/6 & 4/3
\end{bmatrix}
\begin{bmatrix}
c_1 \\ c_2 \\ c_3 \\ c_4
\end{bmatrix}
=
\begin{bmatrix}
0.0 \\ 1.0/h \\ 0.0 \\ 0.0
\end{bmatrix}.
$$

Solving for c_1, c_2, c_3, and c_4 yields $c_1 = -1/(3h)$, $c_2 = 1/(2h)$, $c_3 = 1/h$, and $c_4 = -1/(6h)$. So,

$$
f_x(i) = \frac{-2f_{i-1} - 3f_i + 6f_{i+1} - f_{i+2}}{6\Delta x} + O(\Delta x^3). \tag{5.33}
$$

This approach is systematic to approximate derivatives of function.

Exercise: Approximate the second derivative of a function including f_{i-2}, f_{i-1}, f_i, f_{i+1}, and f_{i+2}.

Note: It is always a good idea to approximate even derivatives (second, fourth, etc.) of a function including the same number of function values from left-hand and right-hand sides of the point of interest, as mentioned before. For instance, it is a good idea to approximate f_{xx} including f_{i-1}, f_i, and f_{i+1} or including f_{i-2}, f_{i-1}, f_i, f_{i+1}, and f_{i+2}. However, it is not appropriate to include, for example, f_{i-2}, f_{i-1}, f_i, and f_{i+1}, in which case the approximation is biased toward the left-hand side.

5.5   Summary of Most Used Derivatives

In Table 5.1, a summary of first, second, third, and fourth derivatives approximation is given for uniform grids.

Table 5.1 is based on central difference approximations. For example, the fourth-order accurate approximation for first derivative is

$$
\left.\frac{df}{dx}\right|_i = \frac{1}{\Delta x}\left[\frac{f_{i-2}}{12} - \frac{2f_{i-1}}{3} + \frac{2f_{i+1}}{3} - \frac{f_{i+2}}{12}\right]. \tag{5.34}
$$

Table 5.1: Central approximation of derivatives. The coefficients of the approximation have to be divided by h, h^2, h^3, and h^4 for the first, second, third, and fourth derivatives, respectively.

Order of derivative	Order of accuracy	-3	-2	-1	0	1	2	3
1st	2	0	0	$-1/2$	0	$1/2$	0	0
	4	0	$1/12$	$-2/3$	0	$2/3$	$-1/12$	0
2nd	2	0	0	1	-2	1	0	0
	4	0	$-1/12$	$4/3$	$-5/2$	$4/3$	$-1/12$	0
3rd	2	0	$-1/2$	1	0	-1	$1/2$	0
	4	$1/8$	-1	$13/8$	0	$-13/8$	1	$-1/8$
4th	2	0	1	-4	6	-4	1	0
	4	$-1/6$	2	$-13/2$	$28/3$	$-13/2$	2	$-1/6$

Note that the weights are equal for the neighbor of f_i at the central point x_i. Also, it is easy to check whether any approximation is wrong, i.e., whether there is any inconsistency. If all the values of f are identical, then its representation is a straight line with zero slope: All the derivatives are zero. Hence, setting the values of f_{i-2}, f_{i-1}, f_i, f_{i+1}, and f_{i+2} to unity (actually, any nonzero constant will do the job) in the approximation (5.34) of the first derivative yields $1/12 - 2/3 + 2/3 - 1/12 = 0$. Also, the straight line has no curvature (hence the second derivative is zero), no inflection (hence the third derivative is zero), etc. For example, the fourth-order accurate approximation for the fourth derivative when f is set to unity reads $-1/6 + 2 - 13/2 + 28/3 - 13/2 + 2 - 1/6 = 0$.

At the boundaries, using higher-order approximations is difficult because the value of the function is not defined outside of the domain. In such conditions, we have to consider the values of the function inside the domain, i.e., nodes from the interior of the domain. Also, in problems where there is information stream from one direction, such as advection terms in fluid flow problems, it is more sound physically (and also more stable numerically, as we will see in Chapter 11) to use the nodes (or at least more nodes) from the upstream direction. Table 5.2 summarizes the biased derivatives, i.e., the derivatives that have to be computed at the left boundary of a domain.

Table 5.2:   Forward approximation of derivatives. The coefficients of the approximation have to be divided by h, h^2, h^3, and h^4 for the first, second, third, and fourth derivatives, respectively.

Order of derivative	Order of accuracy	0	1	2	3	4	5	6
1st	1	-1	1	0	0	0	0	0
	2	$-3/2$	2	$-1/2$	0	0	0	0
	3	$-11/6$	3	$-3/2$	1/3	0	0	0
2nd	2	2	-5	4	-1	0	0	0
	3	35/12	$-26/3$	19/2	$-14/3$	11/12	0	0
3rd	2	$-5/2$	9	-12	7	$-3/2$	0	0
	3	$-17/4$	71/4	$-59/2$	49/2	$-41/4$	7/4	0
4th	2	1	-4	6	-4	1	0	0
	3	35/6	-31	137/2	$-242/3$	107/2	-19	17/6

For example, the first derivative of third-order accurate approximation reads

$$\frac{df}{dx}\bigg|_i = \frac{1}{\Delta x}\left[-\frac{11}{6}f_i + 3f_{i+1} - \frac{3}{2}f_{i+2} + \frac{1}{3}f_{i+3}\right]. \qquad (5.35)$$

Exercise: Approximate the first derivative at the right-hand side boundary of a domain using a third-order accurate approximation.

Exercise: Use central difference approximation for the first order with non-uniform grids. Note that the order of accuracy decreases by one.

5.6   Non-uniform Grids

In most applications, the dependent function has steep gradients at a specific region, while it has small gradients on other parts of the domain. For example, consider heat transfer by convection from a plate, where the fluid temperature changes drastically close to the solid (boundary phenomenon). To accurately estimate the rate of heat transfer, the temperature derivative at the boundary needs to be accurately estimated for both are linked linearly through Fourier law. So, the grid size must be small enough. However, there is no specific need to apply the same small grid size elsewhere. To resolve

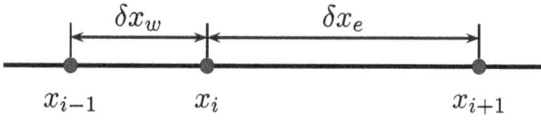

Figure 5.3: Non-uniform grid.

this issue, more nodes need to be clustered only at the boundary. Another example: In calculating shear stress on a body embedded in a fluid flowing around, the velocity gradient is needed at the boundary. Hence, more grid nodes (or control volumes, etc.) need to be clustered at the boundary. In most engineering problems, there are regions with high gradient of dependent function and others region where the dependent function does not change significantly. Using a uniform grid is not recommended because of computation cost and resources. Non-uniform grids yield accurate results compared with uniform grids even when using large number of grid nodes. The bottom line is that we have to figure out how to approximate the derivative of a function for a non-uniform grid.

Forward expansion of a function using Taylor series yields (see Fig. 5.3)

$$f_{i+1} = f_i + \left.\frac{df}{dx}\right|_i (\delta x_e) + \frac{1}{2!}\left.\frac{d^2 f}{dx^2}\right|_i (\delta x_e)^2 + \frac{1}{3!}\left.\frac{d^3 f}{dx^3}\right|_i (\delta x_e)^3 + \cdots .$$

(5.36)

Backward expansion of a function using Taylor series yields

$$f_{i-1} = f_i - \left.\frac{df}{dx}\right|_i (\delta x_w) + \frac{1}{2!}\left.\frac{d^2 f}{dx^2}\right|_i (\delta x_w)^2 - \frac{1}{3!}\left.\frac{d^3 f}{dx^3}\right|_i (\delta x_w)^3 + \cdots ,$$

(5.37)

where δx_e and δx_w are the east and west grid spacings, respectively. In fact, the domain of integration can be mapped by calculating or setting δx before solving the differential equation. Using such a strategy, we can deal with irregular domains as well.

5.6.1   First derivative

The first-order accurate forward scheme is

$$\left.\frac{df}{dx}\right|_i = \frac{f(i+1) - f(i)}{\delta x_e} + O(\Delta x),$$

(5.38)

while the first-order accurate backward scheme is

$$\frac{df}{dx}\bigg|_i = \frac{f(i) - f(i-1)}{\delta x_w} + O(\Delta x). \tag{5.39}$$

Subtracting side by side Eq. (5.37) from Eq. (5.36), after algebraic manipulation, yields

$$\frac{df}{dx}\bigg|_i = \frac{f_{i+1} - f_{i-1}}{\delta x_e + \delta x_w} - \frac{1}{2!}\frac{d^2 f}{dx^2}\bigg|_i (\delta x_e - \delta x_w) + O(\Delta x^2). \tag{5.40}$$

Note that the above equation is not second-order accurate unless δx_e equals δx_w. The leading error is a diffusive term $\left(\frac{d^2 f}{dx^2}\right)$. There is room to improve the accuracy of the scheme by expressing the term $\frac{d^2 f}{dx^2}$ using finite difference, as explained in the following section.

5.6.2   Second derivative

Adding side by side Eqs. (5.36) and (5.37), after algebraic manipulation, yields

$$\frac{d^2 f}{dx^2}\bigg|_i = \frac{2}{(\delta x_e + \delta x_w)}\left[\frac{f_{i+1}}{\delta x_e} - f_i(\frac{1}{\delta x_e} + \frac{1}{\delta x_w}) + \frac{f_{i-1}}{\delta x_w}\right] + O(\Delta x).$$

$$\tag{5.41}$$

The derivative is first-order accurate if $\delta x_e \neq \delta x_w$. It is second-order accurate if $\delta x_e = \delta x_w$.

Note: As said earlier, calculations in engineering take places close to the boundaries: shear stress due to a fluid flow over a (maybe deformable) surface, heat flux exchange between a solid surface and fluid, etc. Any flux (mass, species, momentum, electric charge, heat, etc.) requires calculation of the derivative at the boundaries, in which case a few values of the dependent variable (normal or tangential velocity component, concentration, temperature, etc.) need to be known along the normal to the wall (and close to it). Using a non-uniform grid in order to get a fine meshing close to the wall is not sufficient to get satisfactory results. In fact, it is recommended to use at least second-order accurate derivative with non-uniform grid schemes. Moreover, it is highly advised to test the grid independency of the solution by comparing to half grid size (half interval) solution:

An acceptable solution is ultimately a solution that is practically grid size insensitive.

Example: In the previous paragraph, we mentioned that the most interesting phenomena take place at the boundary. The value of the dependent variable (velocity, temperature, etc.) follows an exponential profile, typically. Let us assume that the function f is as follows:

$$f(x) = e^{-\alpha x}, \tag{5.42}$$

where the value of α defines the gradient of the function at the wall where $x = 0$.

From Table 5.2, the first-order accurate approximation of the derivative at the left-side boundary is

$$f'(x) = \frac{f(x+h) - f(x)}{h} = \frac{f(h) - f(0)}{h}, \tag{5.43}$$

while the second-order accurate approximation is

$$f'(x) = \frac{4f(h) - f(2h) - 3f(0)}{2h}. \tag{5.44}$$

Let us test the value of derivative for $\alpha = 0.1, 1, 2$, using $h = 0.1$ and 0.2. And let us compare the results with exact values.

From Table 5.3, it is obvious that as the gradient (or slope) increases, i.e., large α, the accuracy of the numerical approximation decreases. For instance, for $\alpha = 2.0$ and for $h = 0.2$, the error is 17.57% and 4.0% using the first- and second-order accurate approximations, respectively. However, the error decreases to 9.36% and 1.15% using a smaller interval, say $h = 0.1$.

The lesson is to use less costly, higher order, and small interval near the boundaries to improve the approximation accuracy at reasonable expense.

Table 5.3: Comparison of exact solution with numerical predictions.

	$h = 0.1$				$h = 0.2$		
α	Exact	1st order	2nd order	α	Exact	1st order	2nd order
0.1	−0.1	−0.09950	−0.09999	−0.1	−0.1	−0.09901	−0.09998
1.0	−1.0	−0.95162	−0.99691	−1.0	−1.0	−0.90635	−0.98849
2.0	−2.0	−1.81269	−1.97698	−2.0	−2.0	−1.64840	−1.92012

5.7   Introduction to Numerical Integration

Integration or summation, finding area under a curve, calculating a volume, a surface area of an irregular shape or the length of a path are all examples where we carry out integration. For example, in testing an engine, it is usual to measure work done by the engine by integrating pressure change with differential volume, i.e.,

$$W = -\int p \, dV, \tag{5.45}$$

which is done by plotting pressure, p, against volume, V. The area of the closed curve represents the net work done by the engine, W. However, since the curve is not regular and it is difficult to relate pressure with volume analytically, integration is done numerically. In numerical integration, the area is divided into small segments. Hence, the total area is the sum (integral) of the areas of all those small segments. There are many ways to split up (or approximate) the surface under the curves into surfaces with smaller area.

Figure 5.4 shows three ways to divide the area under the curve (representing the function) extending from $x = a$ to $x = b$. The first and second are usually first-order accurate, and the third one is called trapezoidal method, which is second-order accurate. Table 5.4 summarizes low- and high-order numerical integration used in the literature. Note that the domain of integration is divided into equal segments, h. In the rectangular scheme, the integration or summation is performed by passing a constant function (straight line) between two points. In trapezoidal scheme, the integration is performed by passing a linear function between two points or equivalently by evaluating the function at mid distance between two points. Simpson's $1/3$ scheme is performed by passing a second-order polynomial through three equally spaced points. The Simpson's scheme can be applied over the entire domain of integration as (n even)

$$I = \frac{h}{3}(f_0 + 4f_1 + 2f_2 + 4f_3 + 2f_4 + 4f_5 + \cdots + 2f_{n-2} + 4f_{n-1} + f_n). \tag{5.46}$$

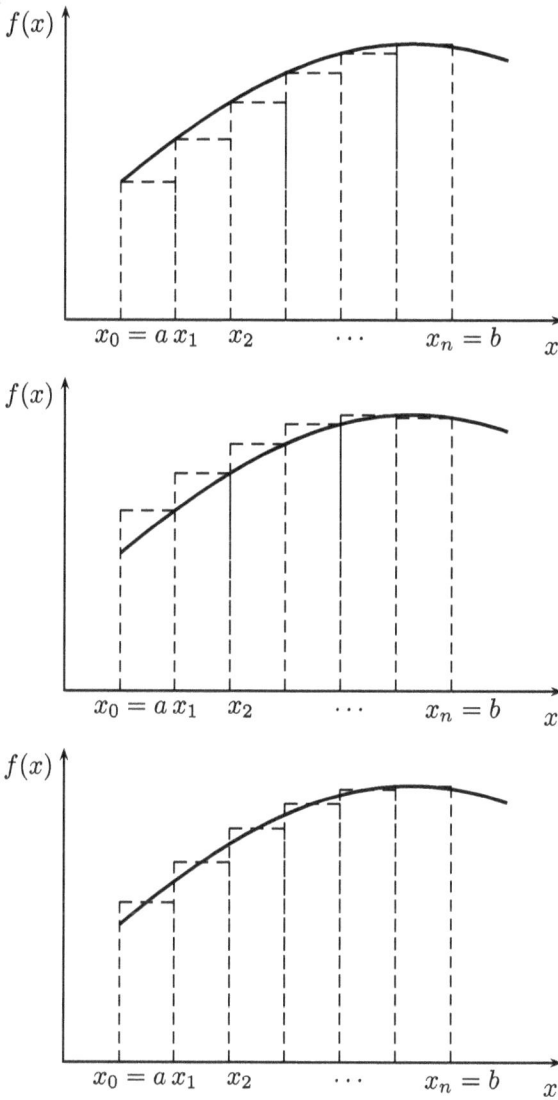

Figure 5.4: Some of the ways to approximate the area of a surface beneath a given curve.

In general, all numerical methods are based on approximating the integral

$$I = \int_a^b f(x)dx \qquad (5.47)$$

Table 5.4:   Numerical integration schemes for one segment of width h.
For the whole integral, the sum over all segments is carried out.

Rectangle	$I \approx h f(x_i)$	1st order
Trapezoid	$I \approx \dfrac{h}{2}\left(f(x_i) + f(x_{i+1})\right)$	2nd order
Simpson	$I \approx \dfrac{h}{3}\left(f(x_i) + 4f(x_{i+1}) + f(x_{i+2})\right)$	4th order

by

$$I \approx \sum_{i=1}^{N} w_i f(x_i), \qquad (5.48)$$

where w_i is a weighting function that depends on the method of integration. $f(x_i)$ is the value of function f at x_i. In other words, the integration is usually converted into polynomial evaluation.

It is possible to use Taylor series approximation for the function and perform the integration as we discussed in Problem 5 in p. 15 of Chapter 1.

For example, consider the following integral:

$$I = \int_0^1 x^2 e^x \, dx. \qquad (5.49)$$

Taylor series expansion of e^x is

$$e^x = 1 + x + \frac{x^2}{2} + \frac{x^3}{3!} + \cdots. \qquad (5.50)$$

Then, the integral can be approximated as

$$I \approx \int_0^1 (x^2 + x^3 + \frac{x^4}{2} + \frac{x^5}{3!} + \cdots) dx. \qquad (5.51)$$

A caveat must be mentioned. Convergence of the series needs to be ensured. The above approximation is valid for small x. So, in general, Taylor series method is not recommended unless the convergence of the series is guaranteed.

Example: Use the trapezoidal and Simpson's methods to perform the previously mentioned integral by using $\Delta x = 0.5$ and 0.2.

For $\Delta x = 0.5$, the trapezoidal method reads

$$I \approx \frac{\Delta x}{2} \left(f(0) + 2f(0.5) + f(1.0) \right)$$

$$= 0.25 \times (0 + 2 \times 0.41218 + 2.71828) = 0.88566. \qquad (5.52)$$

The exact value is 0.718282. Hence, the error is 23.3%.
Simpson's method reads

$$I \approx \frac{\Delta x}{3} \left(f(0) + 4f(0.5) + f(1.0) \right) = 0.727833. \qquad (5.53)$$

Hence, the error is about 1.33%.

For $\Delta x = 0.2$, the trapezoidal method reads

$$I \approx \frac{\Delta x}{2} \left(f(0) + 2 \times [f(0.2) + f(0.4) + f(0.6) + f(0.8)] + f(1.0) \right). \qquad (5.54)$$

Hence,

$$I \approx 0.1 \times (0 + 2 \times [0.048856 + 0.238692$$
$$+ 0.655963 + 1.424346] + 2.718282)$$
$$= 0.7454.$$

The exact value is 0.718282. Hence, the error associated with using $\Delta x = 0.2$ is about 3.77%.

For Simpson's method, we use $\Delta x = 0.25$ because the number of segments should be even:

$$I \approx \frac{\Delta x}{3} \left(f(0) + 4 \times [f(0.25) + f(0.75)] + 2f(0.5) + f(1.0) \right)$$

$$= 0.083333 \times (0 + 4 \times 1.271064 + 2 \times 0.41218 + 2.718282)$$

$$= 0.71891. \qquad (5.55)$$

The error is about 0.087%.

5.8 Gauss Quadrature

Gauss quadrature (GQ) method is an elegant and accurate method compared with the above-mentioned methods. In numerical integration methods, the domain is usually divided into equal-length segments. If the domain is divided into n segments, then it is possible to pass an $(n + 1)$ polynomial through these points:

$$I = \int_a^b f(x)dx = \sum_{i=1}^{n} C_i f(x_i), \qquad (5.56)$$

where x_i are the coordinate points, at which value of function $f(x)$ is known, and C_i are weighting factors.

5.8.1 Increased flexibility

In GQ method, we select both x_i and C_i such that the integral is exact for any polynomial of degree $2n - 1$ or less. To generalize GQ, the function needs to be shifted. It is possible to linearly shift a function from $a \leqslant x \leqslant b$ to $-1 \leqslant t \leqslant 1$ by using a linear mapping:

$$x = \frac{b - a}{2}t + \frac{b + a}{2}. \qquad (5.57)$$

Note that for $x = a$, $t = -1$ and for $x = b$, $t = 1$. Also, $dx = \frac{b-a}{2}dt$. Therefore,

$$I = \int_a^b f(x)dx = \frac{b - a}{2} \int_{-1}^1 f(t)dt = \frac{b - a}{2} \int_{-1}^1 f\left(\frac{b - a}{2}t + \frac{b + a}{2}\right)dt. \qquad (5.58)$$

Let us consider two-point GQ ($n = 2$). The integral, I, can be approximated in domain t as

$$I \approx C_1 f(t_1) + C_2 f(t_2), \qquad (5.59)$$

where C_1, t_1, C_2, and t_2 are unknowns. Therefore, four equations are needed in order to determine those unknowns.

If $f(t) = 1$ (constant function), then

$$I = \int_{-1}^1 1dt = 2 = C_1 + C_2. \qquad (5.60)$$

If $f(t) = t$ (linear function), then

$$I = \int_{-1}^1 tdt = \left.\frac{t^2}{2}\right|_{-1}^1 = 0 = C_1 t_1 + C_2 t_2. \qquad (5.61)$$

If $f(t) = t^2$, then

$$I = \int_{-1}^{1} t^2 dt = \frac{t^3}{3}\Big|_{-1}^{1} = \frac{2}{3} = C_1 t_1^2 + C_2 t_2^2. \tag{5.62}$$

The final relationship can be constructed by assuming $f(t) = t^3$, hence

$$I = \int_{-1}^{1} t^3 dt = \frac{t^4}{4}\Big|_{-1}^{1} = 0 = C_1 t_1^3 + C_2 t_2^3. \tag{5.63}$$

Solving for C_1, C_2, t_1, and t_2 yields $C_1 = C_2 = 1$, $t_1 = -\frac{1}{\sqrt{3}}$, and $t_2 = \frac{1}{\sqrt{3}}$. Hence,

$$I = \int_{-1}^{1} f(t)dt = f\left(-\frac{1}{\sqrt{3}}\right) + f\left(\frac{1}{\sqrt{3}}\right). \tag{5.64}$$

The final result is

$$I = \frac{b-a}{2}\left[f\left(\frac{-1}{\sqrt{3}}m + c\right) + f\left(\frac{1}{\sqrt{3}}m + c\right)\right], \tag{5.65}$$

where $m = (b-a)/2$ and $c = (b+a)/2$. The above scheme is third-order accurate. Fifth-order GQ analysis yields $t_1 = -\sqrt{0.6}$, $t_2 = 0$, $t_3 = \sqrt{0.6}$, $C_1 = 5/9$, $C_2 = 8/9$, and $C_3 = 5/9$.

In general, we can formulate GQ as

$$I = \frac{b-a}{2}\sum_{i=1}^{N} C_i f_i\left(\frac{b-a}{2}t_i + \frac{b+a}{2}\right). \tag{5.66}$$

Example: Integrate the following:

$$I = \int_{-1}^{2} e^{-2x}dx. \tag{5.67}$$

Solution: $\frac{b-a}{2} = \frac{2-(-1)}{2} = 1.5$ and $\frac{b+a}{2} = \frac{2-1}{2} = 0.5$. Hence,

$$e^{-2x} \Rightarrow e^{-2(1.5t+0.5)} \text{ and}$$

$$I = \frac{b-a}{2}\left[f\left(-\frac{1}{\sqrt{t}}\right) + f\left(\frac{1}{\sqrt{t}}\right)\right]$$

$$= 1.5\left[e^{-2(-\frac{1.5}{\sqrt{3}}+0.5)} + e^{-2(\frac{1.5}{\sqrt{3}}+0.5)}\right] = 3.21664.$$

The exact value is 3.6853. The difference is 12.72%. Using higher-order scheme, we get

$$I = \frac{b-a}{2}\left[\frac{5}{9}f\left(-\sqrt{0.6}\right) + \frac{8}{9}f(0) + \frac{5}{9}f\left(\sqrt{0.6}\right)\right].$$

$$I = 1.5\left[\frac{5}{9}\left(e^{-2(-1.5\times\sqrt{0.6}+0.5)} + \frac{8}{9}e^{-2(0+0.5)} + \frac{5}{9}e^{-2(1.5\times\sqrt{0.6}+0.5)}\right)\right]$$

$$= 3.6518. \tag{5.68}$$

The difference between exact and approximate is less than 1%.

5.9　Monte Carlo Method

Monte Carlo method is a statistical method based on many trials of function evaluation. The first step of function $f(x)$ integration, from $x = a$ to $x = b$, is to scale the domain $a \leqslant x \leqslant b$ through an independent variable t, $0 \leqslant t \leqslant 1$ as we did with GQ method. So,

$$x = a + (b-a)t \quad \text{and} \quad dx = (b-a)dt. \tag{5.69}$$

Therefore,

$$\int_a^b f(x)dx = (b-a)\int_0^1 f\left([b-a]t + a\right)dt. \tag{5.70}$$

Hence,

$$I = \frac{(b-a)}{N}\sum_{i=1}^{N} f\left([b-a]t_i + a\right), \tag{5.71}$$

where N is the number of trials and t_i is a random number between 0 and 1.

For instance, if the function $f(x) = 2x^2 + 3e^{-0.5x}$, $a = 1$, and $b = 3.0$, then $f(t) = 2(2t+1)^2 + 3e^{-0.5(2t+1)}$.

The function $f(t)$ is evaluated many times, each time by using a random number between 0 and 1 (the value of t). The process is repeated with many trials. The integral $f(t)$ is the sum of all trials divided by the number of the trials. In fact, at each trial, we are calculating the area of the rectangle with unit base and height equal to the value of the function. The average area of all trial rectangles leads to the area under the curve representing $f(t)$. Hence, the error decreases as the number of trials increases (100000 or more).

5.10   Error and Extrapolation

It is important to know the order of accuracy of the numerical scheme to estimate the accuracy of the scheme and extrapolate the solution to higher accuracy. The accuracy of the solution depends on the segment length h. The exact value of a function, f, can be represented as

$$f_{\text{exact}} = f(h) + Ah^n + O(h^{n+m}), \tag{5.72}$$

where A is a coefficient, n is the order of leading error, and m is the increment in the order of the following error term.

The solution can be carried out for a different segment length, say h/R, where R is any value greater than unity. For instance, if the domain is divided into two smaller segments, i.e., $R = 2$, then the exact solution can be represented as

$$f_{\text{exact}} = f(h/R) + A(h/R)^n + O(h^{n+m}). \tag{5.73}$$

Subtracting the above equations yields

$$0 = f(h) - f(h/R) + Ah^n - A(h/R)^n + O(h^{n+m}). \tag{5.74}$$

Solving for Ah^n yields

$$\text{Error}(h) = Ah^n = \frac{R^n}{R^n - 1} \left[f(h/R) - f(h) \right]. \tag{5.75}$$

Also, solving for $A(h/R)$ yields

$$A(h/R)^n = \frac{1}{R^n - 1} \left[f(h/R) - f(h) \right]. \tag{5.76}$$

Hence, the extrapolated solution to one order higher than the original scheme is

$$f_{\text{extapolated}} = f(h/R) + \frac{1}{R^n - 1} \left[f(h/R) - f(h) \right] + O(h^{n+m}). \tag{5.77}$$

The extrapolation method introduced can be equally valid for integration and/or differentiation of a function.

Table 5.5: Comparison of trapezoidal method and Richardson's extrapolation using two trapezoidal results.

Trapezoidal method ($\Delta x = 1$)	Trapezoidal method ($\Delta x = 0.5$)	Richardson's interpolation	Exact solution
10.107337	8.827116	8.400375	8.389056

For example, the mentioned method can be applied to second-order trapezoidal rule by carrying out the integration twice, with h and $h/2$, then extrapolating the results as mentioned. Since the trapezoidal rule is second-order accurate, $R = 2$, the results of extrapolation will be fourth-order accurate. The procedure can be repeated to obtain higher-order integration. The mentioned extrapolation method is called Richardson's extrapolation.

Example: Evaluate the following integral:

$$I = \int_0^2 xe^x dx, \tag{5.78}$$

using trapezoidal rule (set $\Delta x = 1$, then set $\Delta x = 0.5$). Then, use Richardson's extrapolation and compare the prediction with analytical solution which reads $e^2 + 1$.

If we note I_{Trap_1} and I_{Trap_2}, the approximation of the integral using $\Delta x = 1$ and $\Delta x = 0.5$, respectively, then from (5.77), Richardson's extrapolation I_{Rich} reads (with $R = 2$)

$$I_{\text{Rich}} = I_{\text{Trap}_2} + \frac{1}{2^2 - 1} \left(I_{\text{Trap}_2} - I_{\text{Trap}_1} \right). \tag{5.79}$$

Table 5.5 gives the results. While using no more information on the function to be integrated, it is obvious that the result given by Richardson's extrapolation is much more accurate than the trapezoidal even with $\Delta x = 0.5$.

5.11 Problems

Problem 1: A sheet of metal is intended to be used for a solar collector. The metal sheet has to be corrugated as a sine wave. Therefore, bonding two sheets forms a series of channels for water to flow inside.

It is intended to make 20 channels, each 10 cm in width. The amplitude of the sine function is set to 5 cm. Hence, the shape of each corrugated sheet follows the law, $f(x) = 0.05 \sin(\frac{\pi x}{0.1})$. The differential length is $\sqrt{df^2 + dx^2}$ or

$$dl = \sqrt{1 + \left(\frac{df}{dx}\right)^2} \, dx. \tag{5.80}$$

Hence, the length of one segment is $\int_{x=0}^{10} \sqrt{1 + \left(\frac{df}{dx}\right)^2} \, dx$. Calculate the total length of the sheet by summing the length of each segment.

Problem 2: A wheel of radius $a = 2$ cm rolls without slipping over a cylinder with the same radius. The trajectory of any point on the wheel is known as a cardioid curve, see Fig. 5.5, that is described in polar coordinates by

$$r = 2a(1 + \cos\theta). \tag{5.81}$$

Considering that the differential length in polar coordinates is given by $dl = \sqrt{r^2 + \left(\frac{dr}{d\theta}\right)^2} \, d\theta$, calculate, in cm, the length of the trajectory when the wheel makes one complete revolution about the cylinder.

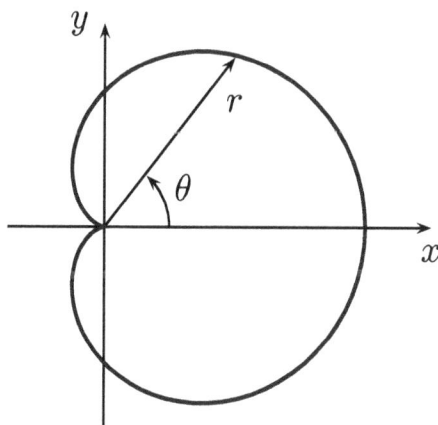

Figure 5.5: Cardioid.

Problem 3: A famous soccer trophy can be represented by a paraboloid that is described in cylindrical coordinates by

$$r = a\,z^2, \qquad (5.82)$$

where $a = 5 \times 10^{-3}\,\mathrm{cm}^{-1}$ and $0 \leqslant z \leqslant 50\,\mathrm{cm}$. The differential area of any surface of revolution is in cylindrical coordinate given by $ds = 2\pi r \sqrt{1 + \left(\frac{dr}{dz}\right)^2}\,dz$.

1. Calculate, in cm^2, the entire area of the circumferential surface of this trophy.
2. Assuming the area density (mass thickness) of the circumferential surface to be constant, calculate the z-coordinate of its center of mass. (*Hint*: Calculate the integral $\int z\,ds$, then divide it by the circumferential surface area.)

Problem 4: A balance spring is a torsion spring for which the oscillation frequency can be finely tuned to get an accurate (and maybe very expensive) watch. It is Archimedean spiral–shaped, which is defined by $r = a\theta$, see Fig. 5.6. By taking $a = 10\,\mathrm{cm}$ and using the expression of the differential length in polar coordinates given in Problem 2, calculate, in cm, the length of the first whorl of the spring starting from the axis, i.e., from the origin of the coordinate system.

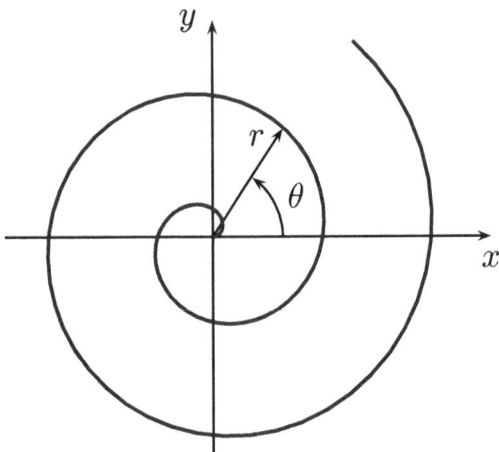

Figure 5.6:   Archimedean spiral.

Chapter 6

Ordinary Differential Equations: Initial Value Problems

6.1 Introduction

There is almost no engineering and/or science topic free of differential equations (ordinary or partial). Whensoever we try to measure a rate of change of any dependent variable with respect to an independent variable, the effort ends up with a differential equation or a system of differential equations, either linear or nonlinear. It may happen, for domains with regular shape, that low-order linear equations can be solved analytically. However, an overwhelming number of problems we face in engineering and science are nonlinear. For simple and exceptional cases, it is possible to find particular analytical solutions for nonlinear equations. In fact, most of them are either difficult or impossible to solve analytically. Moreover, it is difficult to solve both linear and nonlinear problems in irregular domains. Therefore, numerical methods are the only way to follow. For example, hitting a golf ball with an initial spin, the trajectory of the ball can be modeled by Newton's second law of motion, i.e.,

$$m\frac{d^2\mathbf{s}}{dt^2} = \mathbf{F}_{\text{gravity}} + \mathbf{F}_{\text{air, drag}} + \mathbf{F}_{\text{Magnus}}, \qquad (6.1)$$

where m, $\mathbf{s}$, and $\mathbf{F}$ are the mass of the object, the path, and the applied force (gravity, air resistance, and spinning force), respectively. For a motion in a plane, the above vector equations can be written as follows: y-component:

$$\frac{d^2y}{dt^2} = -g - \frac{\rho A\,V}{2m}(C_D v - C_L u) \tag{6.2}$$

and x-component:

$$\frac{d^2x}{dt^2} = -\frac{\rho A\,V}{2m}(C_D u + C_L v), \tag{6.3}$$

where ρ, A, C_D, C_L, u, and v are the air density, cross-sectional area of the ball, drag coefficient, lift coefficient, and x- and y-velocity components, respectively.

In general, differential equations can be classified into two categories: **initial value and boundary value problems.** In the initial value problems, the information propagates (advances, advects, or diffuses) from the initial point(s) toward the domain of the solution at a certain speed. In the boundary value problems, the information instantly propagates form the boundaries into the domain at an infinite speed. In another words, the boundary conditions dictate at any time the domain solution.

6.2   Initial Value Problems

The first-order, initial value ordinary differential equation (ODE) is,

$$\frac{d\phi}{dt} = F(t, \phi). \tag{6.4}$$

In general, F is function of the independent, t, and dependent, ϕ, variables. Higher-order differential equations can be reduced to a system of first-order equations. For instance,

$$\frac{d^2S}{dt^2} + a\left(\frac{dS}{dt}\right)^2 + b\,S = c \tag{6.5}$$

can be cast into two first-order ODEs. Let $\frac{dS}{dt} = \varphi$, where φ is another function. Hence, the second-order ODE becomes

$$\begin{cases} \dfrac{dS}{dt} = \varphi, \\[2mm] \dfrac{d\varphi}{dt} = c - a\varphi^2 - b\,S. \end{cases} \tag{6.6}$$

Therefore, any method of solution used to solve a first-order ODE can equally be applied to solve higher-order ODEs by considering the dependent variable like a vector.

6.3 General Form of Initial Value, First-Order Ordinary Differential Equations

The general form for initial value ODEs is Eq. (6.4). The latter can be linearized using Taylor series about (t_0, ϕ_0) as

$$F(t, \phi) = F(t_0, \phi_0) + \left.\frac{\partial F}{\partial t}\right|_{t_0,\phi_0} (t - t_0) + \left.\frac{\partial F}{\partial \phi}\right|_{t_0,\phi_0} (\phi - \phi_0)$$

$$+ O(\Delta t)^2 + O(\Delta \phi)^2. \tag{6.7}$$

Since $F(t_0, \phi_0)$, t_0, ϕ_0, and derivative of F with respect to t and ϕ are known or can be determined, the above equation can be cast as

$$\frac{d\phi}{dt} + \alpha\phi = f(t), \tag{6.8}$$

where α is a constant and equal to $-\left.\frac{\partial F}{\partial \phi}\right|_{t_0,\phi_0}$ and $f(t)$ is

$$f(t) = \left.\frac{\partial F}{\partial t}\right|_{t_0,\phi_0} t + F(t_0, \phi_0) - \left.\frac{\partial F}{\partial t}\right|_{t_0,\phi_0} t_0 - \left.\frac{\partial F}{\partial \phi}\right|_{t_0,\phi_0} \phi_0, \tag{6.9}$$

which can be written as

$$f(t) = c_0 + c_1 t, \tag{6.10}$$

where $c_0 = F(t_0, \phi_0) - \frac{\partial F}{\partial t} t_0 - \frac{\partial F}{\partial \phi} \phi_0$, that is, a constant. $c_1 = \frac{\partial F}{\partial t}$ evaluated at $t = 0$ is a constant too.

So, Eq. (6.8) can also be written as

$$\frac{d\phi}{dt} + \alpha\phi = c_0 + c_1 t. \tag{6.11}$$

Equation (6.11) can be solved analytically by making the ansatz

$$\phi = \phi_h + \phi_p, \tag{6.12}$$

where ϕ_h is the homogeneous solution and ϕ_p a particular solution. The homogeneous solution is the solution to Eq. (6.11), where the right-hand side is set as zero, i.e.,

$$\frac{d\phi_h}{dt} + \alpha\phi_h = 0. \tag{6.13}$$

This solution reads

$$\phi_h(t) = \phi_0 e^{-\alpha t}, \tag{6.14}$$

where ϕ_0 is the initial value of ϕ at $t = 0$. The particular solution is the solution retrieved by assuming that $\phi_p(t) = a + bt + ct^2$, where a, b, and c are some constants. By introducing the expression of ϕ_p into Eq. (6.11) and identifying a, b, and c, we get

$$\phi_p(t) = \frac{c_0}{\alpha} - \frac{c_1}{\alpha^2} + \frac{c_1}{\alpha}t \tag{6.15}$$

and ultimately retrieve the local solution,

$$\phi(t) = \phi_0 e^{-\alpha t} + \frac{c_0}{\alpha} - \frac{c_1}{\alpha^2} + \frac{c_1}{\alpha}t. \tag{6.16}$$

In the following sections, the linearized equation will be solved using different numerical methods.

6.4 Semi-analytical Method

Solution (6.16) can be used to build step by step the solution over the entire domain by dividing the latter into segments of Δt, say $0, 1, 2, 3, \ldots, i-2, i-1, i, i+1, i+2, \ldots, N$. Points are denoted by $t_0, t_1, \ldots, t_i, \ldots, t_N$. Equation (6.16) can be written for any segment $[t_i, t_{i+1}]$ as

$$\phi_{i+1} = \phi_i e^{-\alpha_i \Delta t} + \frac{c_{0,i}}{\alpha_i} - \frac{c_{1,i}}{\alpha_i^2} + \frac{c_{1,i}}{\alpha_i}\Delta t. \tag{6.17}$$

At each step, the values of $c_{0,i}$, $c_{1,i}$, and α_i need to be computed at point i, where the initial condition considered here is the solution $\phi_i(t = t_i)$ calculated from the previous step solution.

Example: To illustrate the method, let us consider the following differential equation:

$$\frac{dy}{dt} = -y^2 t, \tag{6.18}$$

with initial condition $y = 10$ at $t = 0$. The exact solution is

$$y = \frac{1}{0.5t^2 + 0.1}.$$ (6.19)

The right-hand side, F, of the equation given by

$$F(t, y) = y^2 t$$ (6.20)

is linearized as

$$F(t, y) = y_0^2 t_0 + y_0^2 (t - t_0) + 2y_0 t_0 (y - y_0),$$ (6.21)

where $\alpha = -2y_0 t_0$. The constant term is $c_o = -2y_0^2 t_0$. Hence, the linearized equation can be written as

$$\frac{dy}{dt} - 2y_0 t_0 y = -2y_0^2 t_0$$ (6.22)

and so on. However, this method is not efficient and is difficult to apply to most practical applications.

6.5   Taylor Series Method

Taylor series can be used to solve ODEs, for example

$$\frac{d\phi}{dt} = F(t, \phi),$$ (6.23)

where $F(t, \phi) = 2\phi^2 - 4t\phi$. At $t = t_0$, $\phi = \phi_0$. By using Taylor series,

$$\phi(t) = \phi(t_0) + \left.\frac{d\phi}{dt}\right|_{t_0} (t - t_0) + \left.\frac{d^2\phi}{dt^2}\right|_{t_0} (t - t_0)^2/2 + \cdots,$$ (6.24)

where

$$\begin{aligned}
\frac{d^2\phi}{dt^2} &= \frac{d}{dt}\left(\frac{d\phi}{dt}\right) \\
&= \frac{dF}{dt} = \frac{\partial F}{\partial t} + \frac{\partial F}{\partial \phi}\frac{d\phi}{dt} \\
&= \frac{\partial F}{\partial t} + \frac{\partial F}{\partial \phi}F.
\end{aligned}$$ (6.25)

(Here, we used the definition of the total derivative of a two-independent variable function, $\frac{dF}{dt} = \frac{\partial F}{\partial t} + \frac{\partial F}{\partial \phi}\frac{d\phi}{dt}$.) Hence,

$$\frac{d^2\phi}{dt^2} = -4\phi + (4\phi - 4t)(2\phi^2 - 4t\phi). \tag{6.26}$$

Substituting into Eq. (6.26),

$$\phi(t) = \phi_0 + (2\phi_0^2 - 4t_0\phi_0)(t - t_0) + [-4\phi_0 + (4\phi_0 - 4t_0)$$
$$\times (2\phi_0^2 - 4t_0\phi_0)](t - t_0)^2/2 + \cdots. \tag{6.27}$$

The advantage of the Taylor series method is that we can obtain any order of accuracy by considering more terms. However, it needs the derivatives of the function which get quickly cumbersome and cost-ineffective.

6.5.1 Euler method

The domain of solution is divided into equal-length segments Δx, or h, where x is the independent variable. Note that it is not necessary that the length of each segment be equal. However, for simplicity and illustration, we assume that Δx (or h) is constant. Euler method is no more than using first-order Taylor series:

$$\phi_{i+1} = \phi_i + \frac{d\phi}{dx}\Delta x + \frac{1}{2}\frac{d^2\phi}{dx^2}\Delta x^2 + \cdots. \tag{6.28}$$

Since $\frac{d\phi}{dx} = F$, then

$$\phi_{i+1} = \phi_i + F\Delta x + O(\Delta x^2). \tag{6.29}$$

The above equation is second-order accurate locally. However, after carrying out the procedure many times, the accuracy decreases to first order, which is called global accuracy. Why?

For ϕ_{i+2},

$$\phi_{i+2} = \phi_{i+1} + \frac{d\phi}{dx}\Delta x + \frac{1}{2}\frac{d^2\phi}{dx^2}\Delta x^2 + \cdots. \tag{6.30}$$

Repeating the above procedure for successive points leads to the following:

$$\phi_N = \phi_0 + \sum_{i=0}^{N-1} F_i\Delta x + \frac{1}{2}\frac{d^2\phi}{dx^2}N\Delta x^2. \tag{6.31}$$

But $N = \frac{x_N - x_0}{\Delta x}$. So, substituting into the above equation,

$$\phi_N = \phi_0 + \sum_{i=0}^{N-1} F_i \Delta x + \frac{1}{2} \frac{d^2\phi}{dx^2}(x_N - x_0)\Delta x. \qquad (6.32)$$

Hence, the order of accuracy is decreased by one indeed.

Equation (6.29) is called **explicit (or forward) Euler method** if F is evaluated at point i and **implicit (or backward) Euler method** if F is evaluated at point $i + 1$. The accuracy increases if F is evaluated as the average value of F_i and F_{i+1}. This is the **central Euler method** or **Crank–Nicolson method**. We discuss those options in the following sections.

To illustrate the concept of explicit, implicit, and central approximations, let us solve the following equation numerically:

$$\frac{dy}{dx} = -0.5y, \qquad (6.33)$$

subjected to $y = 2$ at $x = 0$. The reader may argue that the problem is very simple, and the analytical solution is $y = 2e^{-0.5x}$. However, the procedure of the solution is the same regardless of the problem complexity being simple or difficult. Moreover, the feature of each scheme can be illustrated, which is also valid for any problem using the same scheme. The domain needs to be divided into segments of length Δx. We are seeking the value of y at $x = 10$, for example.

6.5.2 Explicit Euler method

Explicit Euler method applied to Eq. (6.33) reads

$$y_{i+1} = y_i - 0.5y_i \Delta x = y_i(1 - 0.5\Delta x). \qquad (6.34)$$

It is very clear that there is a problem if we select Δx such that $0.5\Delta x > 1.0$, the coefficient of y_i on the right hand side will oscillate between negative and positive values of y, i.e., the solution diverges. y_i is bounded only if $0.5\Delta x \leqslant 1.0$. The scheme applied to Eq. (6.33) is said to be **conditionally stable**. It is worth mentioning that the stability of a scheme does not guarantee its accuracy; stability deals only with the numerical solution boundedness in the case the exact solution is itself supposed to be bounded.

The scheme is explicit because the slope, $d\phi/dx$, is evaluated at point x_i. The scheme is second-order accurate locally and first-order accurate globally and conditionally stable. The reader should try to use Δx of 1, 2, and 3, for example.

6.5.3 Implicit Euler method

Implicit Euler method applied to Eq. (6.33) reads

$$y_{i+1} = y_i - 0.5y_{i+1}\Delta x, \qquad (6.35)$$

which can be cast as

$$y_{i+1} = \frac{y_i}{1 + 0.5\Delta x}. \qquad (6.36)$$

It is clear that the scheme is unconditionally stable, i.e., the solution is bounded (and decreasing, for $1/(1+0.5\Delta x) < 1$) regardless of the value of Δx. We can select any value of Δx without any problem as far as the stability is concerned. Note the accuracy of the solution depends on Δx. The scheme is implicit because the slope $d\phi/dx$ is evaluated at point x_{i+1}. Also, the scheme is second-order accurate locally and first-order accurate globally. Note that the main problem with implicit method is that the problem becomes nonlinear if F is a nonlinear function of y. The separation of variable is not possible. Hence, there is a need to solve at each step a nonlinear algebraic equation, using one of the methods described in Chapter 4.

6.5.4 Central scheme

Central Euler method applied to Eq. (6.33) reads

$$y_{i+1} = y_i - 0.5\frac{y_i + y_{i+1}}{2}\Delta x, \qquad (6.37)$$

which can be cast as

$$y_{i+1} = \frac{y_i(1 - 0.5\Delta x/2)}{1 + 0.5\Delta x/2}. \qquad (6.38)$$

It is clear that the scheme is conditionally stable. However, its stability condition is better than explicit scheme (it is possible to select double segment size, Δx, for central scheme (CS) compared with explicit scheme because in CS, Δx is divided to 2). Moreover, the scheme is third-order accurate locally and second-order accurate globally because the slope $d\phi/dx$ is evaluated as the average of the slopes at x_i and x_{i+1}. To prove this, we use forward Taylor expansion for y_{i+1} at $x_{i+1/2}$:

$$y_{i+1} = y_{i+1/2} + \frac{dy}{dx}\bigg|_{i+1/2} \frac{\Delta x}{2} + \frac{1}{2}\frac{d^2y}{dx^2}\bigg|_{i+1/2} \left(\frac{\Delta x}{2}\right)^2$$
$$+ \frac{1}{3!}\frac{d^3y}{dx^3}\bigg|_{i+1/2} \left(\frac{\Delta x}{2}\right)^3 + \cdots . \qquad (6.39)$$

And we use backward Taylor expansion for y_i at $x_{i+1/2}$:

$$y_i = y_{i+1/2} - \frac{dy}{dx}\bigg|_{i+1/2} \frac{\Delta x}{2} + \frac{1}{2}\frac{d^2y}{dx^2}\bigg|_{i+1/2} \left(\frac{\Delta x}{2}\right)^2$$
$$- \frac{1}{3!}\frac{d^3y}{dx^3}\bigg|_{i+1/2} \left(\frac{\Delta x}{2}\right)^3 + \cdots . \qquad (6.40)$$

Subtract Eq. (6.40) from Eq. (6.39):

$$y_{i+1} = y_i + \frac{dy}{dx}\bigg|_{i+1/2} \Delta x + \frac{1}{3}\frac{d^3y}{dx^3}\bigg|_{i+1/2} \left(\frac{\Delta x}{2}\right)^3 + \cdots , \qquad (6.41)$$

where $\frac{dy}{dx}\big|_{i+1/2}$ is F is taken as equal to the average of F_i and F_{i+1}. Hence, the local order of the accuracy is third order and global order of accuracy is second order.

The main problem of central scheme is evaluating F at point $i+1$ where the value of the dependent variable is not known. To overcome this shortcoming, it is suggested to split the procedure into two steps (called **predictor and corrector, also known as second-order Runge–Kutta method**). In the first step, explicit scheme is used to determine the value of the dependent variable at point x_{i+1}, which is used to evaluate F at x_{i+1}, called predicted value. In the second step, the central scheme is used as before.

6.6 Higher-Order Methods

We already introduced one of the higher-order accurate methods, the central scheme. In the following section, a few other higher-order methods are introduced.

6.6.1 Taylor series method

Instead of truncating the Taylor series after the second term, why not involve more terms?

$$y_{i+1} = y_i + \frac{dy}{dx}\Delta x + \frac{1}{2}\frac{d^2y}{dx^2}\Delta x^2 + O(\Delta x^3). \tag{6.42}$$

Since $\frac{dy}{dx} = F$, $\frac{d^2y}{dx^2} = \frac{dF}{dx}$ and so on. Hence,

$$y_{+1} = y_i + F\Delta x + \frac{1}{2}\frac{dF}{dx}\Delta x^2 + \cdots. \tag{6.43}$$

Note that, as mentioned earlier, since $F(x, y)$ is function of x and y, then $dF = \frac{\partial F}{\partial x}dx + \frac{\partial F}{\partial y}dy$. Hence, $\frac{dF}{dx} = \frac{\partial F}{\partial x} + \frac{\partial F}{\partial y}\frac{dy}{dx}$.

Accordingly,

$$y_{i+1} = y_i + F\Delta x + \frac{1}{2}\left(\frac{\partial F}{\partial x} + \frac{\partial F}{\partial y}F\right)\Delta x^2 + O(\Delta x^3). \tag{6.44}$$

If we do not want to perform partial differentiation of F with respect of x and y, we can expand, alternatively, F at the middle point $(x_{i+1/2} = x_i + \frac{\Delta x}{2})$ as

$$F\left(x_i + \frac{\Delta x}{2}, y_i + \frac{\Delta x}{2}F(x_i, y_i)\right)$$

$$= F(x_i, y_i) + \frac{\Delta x}{2}\left[\frac{\partial F(x_i, y_i)}{\partial x} + \frac{\partial F(x_i, y_i)}{\partial y}\right] + O(\Delta x^2). \tag{6.45}$$

Therefore,

$$y_{i+1} = y_i + \Delta x F\left(x_i + \Delta x/2, y_i + \frac{\Delta x}{2}F(x_i, y_i)\right) + O(\Delta x^3). \tag{6.46}$$

The above equation can be written in steps:

$$k_1 = \Delta x F(x_i, y_i),$$
$$k_2 = \Delta x F(x_i + \Delta x/2, y_i + k_1/2), \qquad (6.47)$$
$$y_{i+1} = y_i + k_2.$$

The method is $O(\Delta x^3)$ locally and $O(\Delta x^2)$ globally. Also, the method is called second-order Runge–Kutta method.

6.6.2 Fourth-order Runge–Kutta method

One of the most popular methods is the Runge–Kutta (RK) method because of its accuracy and it needs only initial values. It is an explicit method. So, it is (expected to be) conditionally stable, as any other explicit method. It is a multi-step method. Let us consider the following initial value ODE:

$$\frac{dy}{dx} = F(x, y) \quad y(x_0) = y_0. \qquad (6.48)$$

Explicit Euler method is

$$y_{i+1} = y_i + \Delta x F(x_i, y_i). \qquad (6.49)$$

As mentioned, the method is second-order accurate $(O(\Delta x^2))$ locally. To improve the accuracy, we need to involve more terms. This can be done as follows:

$$\frac{dF}{dx} = \frac{\partial F}{\partial x} + \frac{\partial F}{\partial y}\frac{dy}{dx}, \qquad (6.50)$$

i.e.,

$$\frac{dF}{dx} = \frac{\partial F}{\partial x} + \frac{\partial F}{\partial y}F. \qquad (6.51)$$

Using Taylor series with extra terms up to order of $O(\Delta x^3)$,

$$y_{i+1} = y_i + F(x_i, y_i)\Delta x + \frac{1}{2}\left[\frac{\partial F}{\partial x} + \frac{\partial F}{\partial y}F\right]_{x_i, y_i} \Delta x^2 + O(\Delta x^3). \quad (6.52)$$

We can show that the last two terms are the same as expanding function F at $(x_i + 1/2\Delta x)$ and at $(y_i + 1/2\Delta y)$:

$$F\left(x_i + \frac{1}{2}\Delta x, y_i + \frac{1}{2}\Delta y\right)$$

$$= F(x_i, y_i) + \left[\frac{\partial F}{\partial x} + \frac{\partial F}{\partial y}F\right]_{x_i,y_i} \frac{\Delta x}{2} + O(\Delta x^2). \quad (6.53)$$

Substituting Eq. (6.53) into Eq. (6.52) yields

$$y_{i+1} = y_i + \Delta x F\left(x_i + \frac{1}{2}\Delta x + y_i + \frac{1}{2}\Delta y\right) + O(\Delta x^3). \quad (6.54)$$

Hence,

$$k_1 = \Delta x F(x_i, y_i),$$
$$k_2 = \Delta x F\left(x_i + \frac{1}{2}\Delta x, y_i + \frac{1}{2}k_1\right), \quad (6.55)$$
$$y_{i+1} = y_i + k_2,$$

which is called second-order RK method, which is third-order accurate locally and second-order accurate globally. Similarly, we can construct higher-order RK methods.

The most popular method is the fourth-order Runge–Kutta (RK4) method:

$$y_{i+1} = y_i + \frac{1}{6}(k_1 + 2k_2 + 2k_3 + k_4), \quad (6.56)$$

where k_1 is the same step as we did for prediction method for CS:

$$k_1 = F(x_i, y_i)\Delta x \quad (6.57)$$

and

$$k_2 = F(x_i + \Delta x/2, y_i + k_1/2)\Delta x, \quad (6.58)$$
$$k_3 = F(x_i + \Delta x/2, y_i + k_2/2)\Delta x, \quad (6.59)$$
$$k_4 = F(x_i + \Delta x, y_i + k_3)\Delta x. \quad (6.60)$$

The main features of RK4 method are that it is

- explicit,
- fifth-order accurate locally and fourth-order accurate globally, and
- conditionally stable.

If a system of first-order ODEs is considered, then the functions y, F, k_1, k_2, k_3, and k_4 must be treated as vectors.

6.6.3 Polynomial methods

The polynomial methods are another class of methods that involve more points besides y_i and y_{i+1}. They are not self-starting methods. In general, we need to carry out calculations using one of the previously mentioned methods, such as RK method, for a few steps and storing values of those calculations, before switching to the polynomial methods. Note that starting solution with low-order accuracy (first order for example) will reduce the order of accuracy of the entire solution. These methods are based mainly on a polynomial. For example, the fourth-order Adam–Bashforth method is a polynomial that uses and satisfies points y_{i-3}, y_{i-2}, y_{i-1}, and y_i, i.e.,

$$y_{i+1} = y_i + \frac{1}{24}(55y_i - 59y_{i-1} + 37y_{i-2} - 9y_{i-3})\Delta x. \qquad (6.61)$$

Another variation of polynomial method is called Adams–Moulton method:

$$y_{i+1} = y_i + \frac{1}{24}(9y_{i+1} + 19y_i - 5y_{i-1} + y_{i-2})\Delta x. \qquad (6.62)$$

Note that the method is implicit because y_{i+1} is on the right-hand side of the equation. To overcome such a problem, a two-step method can be used (predictor-corrector):

$$y_{i+1}^p = y_i + \frac{1}{24}\left(55y_i - 59y_{i-1} + 37y_{i-2} - 9y_{i-3}\right)\Delta x \qquad (6.63)$$

and

$$y_{i+1} = y_i + \frac{1}{24}\left(9y_{i+1}^p + 19y_i - 5y_{i-1} + y_{i-2}\right)\Delta x. \qquad (6.64)$$

Note that, for example, y_{i+1} means that $F(x_{i+1}, y_{i+1})$ should be evaluated at x_{i+1} and y_{i+1}.

6.6.4   Verlet method

The Verlet method is a numerical scheme mainly used to solve Newton's second law (summation of forces equal to the rate of change of the linear momentum), i.e., for constant mass,

$$\frac{d^2x}{dt^2} = \frac{F}{m}, \tag{6.65}$$

which can be split into two equations:

$$\frac{dx}{dt} = V \quad \text{and} \quad \frac{dV}{dt} = \frac{F}{m}. \tag{6.66}$$

The problem is closed by specifying the initial location and initial velocity of the particle (or body, in general). It is mostly used in molecular dynamics simulations. The method is second-order accurate. The method is based mainly on Taylor series expansion. Hence,

$$x(t + \Delta t) = x(t) + \frac{dx}{dt}\Delta t + \frac{1}{2}\frac{d^2x}{dt^2}\Delta t^2 + \cdots \tag{6.67}$$

and

$$x(t - \Delta t) = x(t) - \frac{dx}{dt}\Delta t + \frac{1}{2}\frac{d^2x}{dt^2}\Delta t^2 + \cdots. \tag{6.68}$$

By adding the above equations, we get

$$x(t + \Delta t) = 2x(t) - x(t - \Delta t) + \frac{d^2x}{dt^2}\Delta t^2. \tag{6.69}$$

Note that initially, i.e., at $t = 0$, no information is available to evaluate $x(t - \Delta t)$. To resolve this issue, we need to use Taylor series expansion:

$$x(0 - \Delta t) = x(t) - \frac{dx}{dt}\Delta t + \frac{1}{2}\frac{d^2x}{dt^2}\Delta t^2 + \cdots. \tag{6.70}$$

This step is needed to start the solution.

The scheme is to solve

$$x_{i+1} = 2x_i - x_{i-1} + \frac{F}{m}\Delta t^2, \tag{6.71}$$

and velocity of the particle can be calculated as

$$V_i = \frac{dx_i}{dt} = \frac{x_{i+1} - x_{i-1}}{2\Delta t} \tag{6.72}$$

or

$$V_{i+1} = V_i + \frac{F_{i+1} + F_i}{2m}\Delta t. \tag{6.73}$$

The algorithm used by molecular dynamics simulation is as follows:

1. Calculate the new position of the particle:

$$x_{i+1} = x_i + V_i\Delta t + \frac{1}{2}\frac{F_i}{m}\Delta t^2. \tag{6.74}$$

2. Calculate the particle velocity at the new location:

$$V_{i+1} = V_i + \frac{F_{i+1} + F_i}{2m}\Delta t. \tag{6.75}$$

The algorithm is second-order accurate and considered semi-implicit.

Exercise: A ball thrown into air at a velocity of 30 m/s at an angle of 60° from horizontal axis. The ball was spinning with $w = 20$ rad/s. The governing equations are

$$\frac{d^2x}{dt^2} = -\frac{a}{m}\sqrt{\dot{x}^2 + \dot{y}^2}\dot{x} - \frac{wb}{m}\dot{y} \tag{6.76}$$

and

$$\frac{d^2y}{dt^2} = -g - \frac{a}{m}\sqrt{\dot{x}^2 + \dot{y}^2}\dot{y} + \frac{wb}{m}\dot{x}, \tag{6.77}$$

where a, b, and m are constants. The gravitational acceleration, g, is 9.81 m/s^2. Set the constants a, b, and m to 0.05, 0.02, and 0.25, respectively. Using Δt of 0.5 and 1.0, plot x, y, $\dot{x}$, and $\dot{y}$ as functions of time up to 120 s. Use second-order Euler, RK4, and Verlet methods, and compare the results.

6.7 Nonlinear ODE

The methods mentioned above can be used to solve nonlinear ODEs. However, the resulting algebraic equations are also nonlinear and consequently are cost-ineffective. One of the methods to overcome such a difficulty is to linearize the ODE locally by using Taylor series expansion. The most general case is the following equation:

$$\frac{dx}{dt} = F(t, x), \tag{6.78}$$

where $F(t, x)$ is nonlinear with respect to x. Using first-order, implicit Euler method, we get

$$x_{i+1} = x_i + F(t_i + \Delta t, x_{i+1})\Delta t. \tag{6.79}$$

Let us expand $F(t, x)$ using Taylor series:

$$F(t + \Delta t, x_{i+1}) = F(x_i, t_i) + \left.\frac{\partial F}{\partial t}\right|_i \Delta t + \left.\frac{\partial F}{\partial x}\right|_i (x_{i+1} - x_i) + \cdots \tag{6.80}$$

and substitute in Eq. (6.79):

$$x_{i+1} = \frac{x_i + F_i\Delta t + \left.\frac{\partial F}{\partial t}\right|_i \Delta t^2 - x_i \left.\frac{\partial F}{\partial x}\right|_i \Delta t}{1 - \left.\frac{\partial F}{\partial x}\right|_i \Delta t}. \tag{6.81}$$

All the terms involved in the right-hand side of Eq. (6.81) are known at step i.

Another way to solve the nonlinear algebraic equation (6.79) is to use Newton–Raphson (NS) method. We can write Eq. (6.79) as

$$x_{i+1} - S = 0, \tag{6.82}$$

where

$$S = x_i + F(t_i + \Delta t, x_{i+1})\Delta t \tag{6.83}$$

is a function of t_i, x_i, and x_{i+1}. The above equation can be solved iteratively by guessing the value of x_{i+1}, see Section 4.3:

$$x_{i+1}^{k+1} = x_{i+1}^k - \frac{S^k}{\left(\frac{\partial S}{\partial x_{i+1}}\right)^k}, \tag{6.84}$$

where superscript k is the iteration number,

$$S^k = x_i + F(t_i + \Delta t, x_{i+1}^k)\Delta t, \qquad (6.85)$$

and

$$\left(\frac{\partial S}{\partial x}\right)^k = \frac{\partial F}{\partial x}\bigg|_{x=x_{i+1}^k} \Delta t. \qquad (6.86)$$

The iterations proceed until the solution converges locally within a specified tolerance.

6.8 Order of Accuracy, Consistency, Stability, and Convergence

Taylor series is a very powerful tool to assist many aspects of the numerical schemes. The most prominent analysis that relies on this series is the local and global consistency. The latter is discussed in the following.

6.8.1 Order of accuracy and consistency

The order of accuracy of a scheme is the lower-order terms in the truncated Taylor series of each term in the scheme, which is called local error. Usually, the global error is one order less than the local error. This point was discussed earlier. On the other hand, and from an intuitive standpoint, consistency ensures that if the size of the independent variable (segment size) approaches zero, such as $\Delta x \to 0$, then the original differential equation is recovered. Most of the widely used numerical schemes are consistent. However, there are a few inconsistent schemes that can be found in the literature. *As a recommendation, never use inconsistent scheme.* The following section introduces a method of assessing the consistency of any numerical scheme.

6.8.2 Consistency analysis

The solution to any differential equation numerically is not exact. There are a few types of errors that are involved in the process of approximating and converting a differential equation into algebraic equations and approximating their solution afterward. However, the topic of this section is to show that the approximated equation is different from the original equation that we are planning to solve. It is essential to understand what kind of problem we will end up with after approximation, which helps us to reduce the error associated with the successive approximations.

To illustrate the concept, let us work on an example. Assume we intend to solve the following equation:

$$\frac{df}{dx} + af = F(x). \tag{6.87}$$

Using the explicit Euler method gives

$$f_{i+1} = f_i - af_ih + hF(t), \tag{6.88}$$

where h is the segment length or Δx.

Expand the left-hand side of the equation using Taylor series:

$$f_{i+1} = f_i + \frac{df}{dx}h + \frac{1}{2!}\frac{d^2f}{dx^2}h^2 + \frac{1}{3!}\frac{d^3f}{dx^3}h^3 + \cdots . \tag{6.89}$$

Substitute the above equation on the left-hand side of the previous equation:

$$f_i + \frac{df}{dx}h + \frac{1}{2!}\frac{d^2f}{dx^2}h^2 + \frac{1}{3!}\frac{d^3f}{dx^3}h^3 + \cdots = f_i - af_ih + hF(x). \tag{6.90}$$

Eliminating f_i and dividing the equation by h yields

$$\frac{df}{dx} + af = F(x) - \frac{1}{2}\frac{d^2f}{dx^2}h - \frac{1}{6}\frac{d^3f}{dx^3}h^2. \tag{6.91}$$

Indeed, we are solving the above equation (which we call modified equation) instead of the original equation. There are extra terms on the right-hand side. However, if h value vanishes (goes to zero), the original equation can be recovered. Hence, the method is **consistent**. Also, the leading error is first order (the power of h is one). Moreover, the leading error coefficient is $-\frac{d^2f}{dx^2}$, which is negative diffusion (i.e., anti-diffusion). Diffusion term always smears the solution.

In other words, it smooths the sharp gradients, whereas anti-diffusion exacerbates them.

Let us use the implicit Euler method for the same problem. Hence,

$$f_{i+1} = f_i - af_{i+1}h + F(x)h. \tag{6.92}$$

Rearranging the equation,

$$(1 + ah)f_{i+1} = f_i + F(x)h. \tag{6.93}$$

Using Taylor series as before to expand f_{i+1},

$$(1 + ah)f_i + (1 + ah)\left(\frac{df}{dx}h + \frac{1}{2!}\frac{d^2 f}{dx^2}h^2 + \frac{1}{3!}\frac{d^3 f}{dx^3}h^3 + \cdots\right)$$
$$= f_i + F(x)h. \tag{6.94}$$

It ends up that we are solving the following equation:

$$\frac{df}{dx} + \frac{a}{1 + ah}f = \frac{F(x)}{1 + ah} - \frac{1}{2}\frac{d^2 f}{dx^2}h - \frac{1}{6}\frac{d^3 f}{dx^3}h^2 + \cdots, \tag{6.95}$$

which is consistent and the leading error is first order.

6.8.3 Stability analysis

We should differentiate between physical stability and numerical stability. The problem is physically stable if the solution is bounded, i.e., the dependent variable asymptotically approaches a certain value as the independent variable increases. If the problem is physically bounded, then the numerical scheme should also produce a bounded solution. There are a few methods developed in the literature to assess the stability of a numerical scheme.

In general, we can express the first-order ODE as

$$\frac{dy}{dx} = F(x, y), \tag{6.96}$$

and the numerical solution for the above equation can be expressed as

$$y_{i+1} = Gy_i, \tag{6.97}$$

where G is the amplification factor, also called gain factor, which can be a real or a complex number. After N number of iterations or updates, the above equation leads to

$$y_N = G^N y_0. \tag{6.98}$$

As $N \to \infty$, the numerical solution is bounded (the scheme is stable) if $|G| \leqslant 1$.

Example: Let us examine the stability of the following equation:

$$\frac{dy}{dx} = \alpha y + F(x). \tag{6.99}$$

Note that the stability of the problem will not be affected by the force term as long as the latter does not depend on the depended variable y (or any of its derivatives). Hence, we set $F(x)$ to zero. Also, note that we cannot study the stability of a physically unstable problem, i.e., $\alpha > 0$. Hence, the problem we are giving consideration to is when $\alpha < 0$, i.e., negative value of α. Let us examine the problem when explicit Euler method is used:

$$y_{i+1} = y_i + \alpha y_i h, \tag{6.100}$$

where h is the segment length. Then,
$y_1 = y_0(1 + \alpha h)$,
$y_2 = y_1(1 + \alpha h) = y_0(1 + \alpha h)^2$,
$y_3 = y_2(1 + \alpha h) = y_0(1 + \alpha h)^3$,
and so on,
$y_N = y_0(1 + \alpha h)^N$. Therefore, $G = (1 + \alpha h)$ for stability condition, $|G| \leqslant 1$. Hence, $|1 + \alpha h| \leqslant 1$.
    Work the same example using implicit Euler method.

6.8.4   Convergence

A scheme applied to a problem is said to be convergent if the solution converges to the exact (analytical) solution as size step of the independent variable (Δx) approaches zero. However, in most cases, we do not know the exact solution. (Hence the purpose of the numerical methods!) For a well-posed problem, i.e., a problem that admits a unique solution in the sense of Hadamard, a theorem established by Lax and Richtmyer (see Section 6.10) may help in this situation.

Without getting into too much detail, the theorem states that for a well-posed problem, the solution given by the scheme converges to the exact solution if the scheme is consistent and stable. This is called Lax–Richtmyer (or Lax equivalence) theorem.

6.9   Examples

To illustrate the concepts presented in this chapter, a few examples are worked out.

Example 1: Taken out of a furnace at 400°C, a body is cooled in room temperature at 20°C. The body's properties are as follows: density: $\rho = 5000$ kg/m^3, specific heat: $c = 800$ J/kg·K, emissivity: $\epsilon = 0.8$, volume: $V = 0.008$ m^3, and surface area: $A = 0.24$ m^2. Determine the cooling curve when: (a) neglecting radiation, (b) including the radiation effect. The governing equation is

$$\rho c \frac{dT}{dt} = -hA(T - T_a) - \epsilon A(T^4 - T_a^4),$$

where $h = 10$ W/m^2.K is the convective heat transfer coefficient. Use central method, implicit Euler, second-order Taylor series, and RK4 methods. For each method, examine the order of accuracy, consistency, and stability. Compare the results using Δt of 0.1, 1, and 10.

Example 2: Lotka–Volterra problem is a system of two nonlinear ODEs describing the evolution of the population of preys, x, and predators, y. The problem reads

$$\begin{cases} \dfrac{dx}{dt} = \alpha x - \beta xy \\[2mm] \dfrac{dy}{dt} = \gamma xy - \delta y \end{cases} \qquad t > 0, \qquad (6.101)$$

where α, β, γ, and δ are some positive constants. Initial conditions are

$$\begin{cases} x(0) = a, \\ y(0) = b, \end{cases} \qquad (6.102)$$

where a and b are positive constants too. This problem is known to be periodic.

We intend to solve this problem using explicit Euler and RK4 methods. Let Δt be the time step, t_i the current time, and (x_i, y_i) the approximation of the solution at this time.

- Explicit Euler method applied to this problem reads

$$x_{i+1} = x_i + \Delta t \left(\alpha x_i - \beta x_i y_i\right),$$
$$y_{i+1} = y_i + \Delta t \left(\gamma x_i y_i - \delta y_i\right),$$

(6.103)

with $x_0 = a$ and $y_0 = b$.
- Explicit RK4 method applied to this problem reads

$$k_{x1} = \Delta t \left(\alpha x_i - \beta x_i y_i\right),$$

$$k_{y1} = \Delta t \left(\gamma x_i y_i - \delta y_i\right),$$

$$k_{x2} = \Delta t \left(\alpha \left[x_i + \frac{1}{2}k_{x1}\right] - \beta \left[x_i + \frac{1}{2}k_{x1}\right]\left[y_i + \frac{1}{2}k_{y1}\right]\right),$$

$$k_{y2} = \Delta t \left(\gamma \left[x_i + \frac{1}{2}k_{x1}\right]\left[y_i + \frac{1}{2}k_{y1}\right] - \delta \left[y_i + \frac{1}{2}k_{y1}\right]\right),$$

$$k_{x3} = \Delta t \left(\alpha \left[x_i + \frac{1}{2}k_{x2}\right] - \beta \left[x_i + \frac{1}{2}k_{x2}\right]\left[y_i + \frac{1}{2}k_{y2}\right]\right),$$

$$k_{y3} = \Delta t \left(\gamma \left[x_i + \frac{1}{2}k_{x2}\right]\left[y_i + \frac{1}{2}k_{y2}\right] - \delta \left[y_i + \frac{1}{2}k_{y2}\right]\right),$$

$$k_{x4} = \Delta t \left(\alpha \left[x_i + k_{x3}\right] - \beta \left[x_i + k_{x3}\right]\left[y_i + k_{y3}\right]\right),$$

$$k_{y4} = \Delta t \left(\gamma \left[x_i + k_{x3}\right]\left[y_i + k_{y3}\right] - \delta \left[y_i + k_{y3}\right]\right),$$

$$x_{i+1} = x_i + \frac{1}{6}\left(k_{x1} + 2k_{x2} + 2k_{x3} + k_{x4}\right),$$

$$y_{i+1} = y_i + \frac{1}{6}\left(k_{y1} + 2k_{y2} + 2k_{y3} + k_{y4}\right),$$

(6.104)

with the same initial conditions.

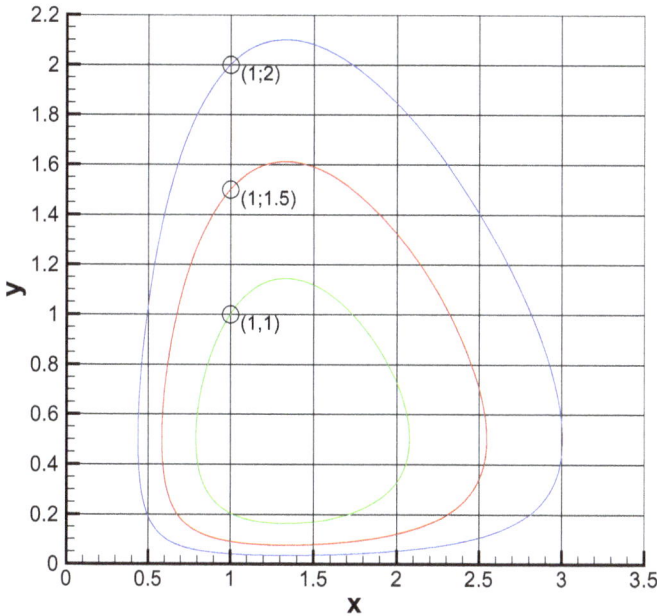

Figure 6.1: Phase diagram computed using RK4 method with time step $\Delta t = 10^{-2}$. For each curve, different starting conditions are used: $(x_0; y_0) = (1; 1)$, $(1; 1.5)$, and $(1; 2)$.

Figure 6.1 shows the phase diagram of the solution obtained using RK4 method with $\Delta t = 10^{-2}$ and the computation is led up to $t = 10$.

Figure 6.2 shows the solution $x(t)$ computed by Euler method (for two time steps $\Delta t = 10^{-2}$ and $\Delta t = 10^{-3}$) and RK4 (for time step $\Delta t = 10^{-2}$). Clearly, Euler method with time step $\Delta t = 10^{-2}$ is not sufficiently accurate as it exhibits a non-periodic solution. One needs to decrease the time step down to $\Delta t = 10^{-3}$ in order to get acceptable results. Conversely, RK4 method is accurate even with time step $\Delta t = 10^{-2}$.

6.10   Problems

Problem 1: In its transient state, Lorenz problem presented in Section 4.8 reduced the atmospheric non-isothermal flow problem

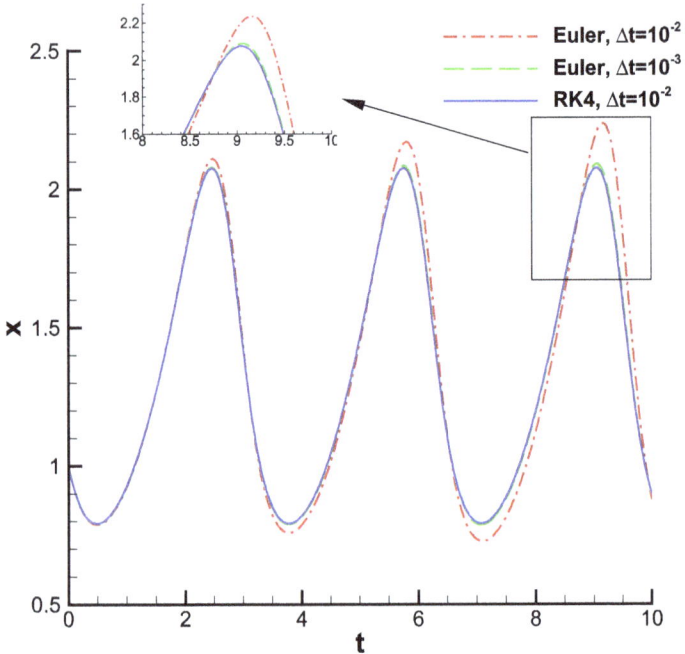

Figure 6.2: Variations of the solution $x(t)$ computed using Euler method (with $\Delta t = 10^{-2}$ and $\Delta t = 10^{-3}$) and RK4 method (with $\Delta t = 10^{-2}$). The inset is a closeup of the larger plot.

to a system of initial value ODEs:

$$\frac{dx}{dt} = 10(y - x), \tag{6.105}$$

$$\frac{dy}{dt} = x(25 - z) - y, \tag{6.106}$$

$$\frac{dz}{dt} = xy - 2z. \tag{6.107}$$

Solve the above equations for initial conditions $x(0) = 1$, $y(0) = 0$, and $z(0) = 20$. Use $\Delta t = 0.2, 0.1$, and 0.05. Compare the results produced using explicit Euler method, modified Euler method (second-order accurate), and RK4 method. Carry out the calculation up to a total time of $t = 10$. Plot x, y, and z versus time. Also, plot x versus z.

Problem 2: The kinetics of a chemical reaction is modeled by a system of coupled first-order differential equations:

$$\frac{dX}{dt} = A - (B+1)X = X^2Y, \tag{6.108}$$

$$\frac{dY}{dt} = BX - X^2Y, \tag{6.109}$$

where X and Y are the concentrations of the chemical species and A and B are constants. For $A = 1$ and $B = 2$, solve the above equations with initial conditions $X = 0.6$ and $Y = 0.4$. Compare the predictions of explicit, implicit, and modified Euler methods using $\Delta t = 0.01$; carry out the calculation up to time $t = 1.0$.

Problem 3: The major components of an electric circuit are capacitor, C, resistor, R, and inductance, L. The capacitor is initially charged to a voltage V_0. Suddenly, the switch is closed at time $t = 0$. The governing equation can be formulated as

$$LC\frac{d^2V}{dt^2} + RC\frac{dV}{dt} = -V. \tag{6.110}$$

Let $C = 2 \times 10^{-6}$ F, $L = 0.5$ H, and $R = 100$ Ohm. At $t = 0$, $V = 100$; V and $dV/dt = 0$ V/s. Use the modified Euler method (prediction-correction) to solve the above equation up to time $t = 60$ s. Use different time steps and check the accuracy of the predicted results by comparing with the analytical solution:

$$V = V_0 \exp\left(-\frac{Rt}{2L}\right)\left[\left(\frac{1}{2} + \frac{R}{4Lb}\right)\exp(bt) + \left(\frac{1}{2} - \frac{R}{4Lb}\right)\exp(-bt)\right], \tag{6.111}$$

where $b = \frac{1}{LC} - \frac{R^2}{4L^2}$.

Project 1: Develop a computer code (using either Matlab or Python) to simulate CO_2 concentration in atmosphere. The following model is taken from the work of J.C.G. Walker. The model simulates the CO_2 reaction of carbonates with alkalinities in shallow and deep ocean waters. The model equations are

$$\frac{dp}{dt} = \frac{p_s - p}{d} + \frac{f(t)}{\mu_1}, \tag{6.112}$$

$$\frac{d\sigma_s}{dt} = \frac{1}{\nu_s}\left[w(\sigma_d - \sigma_s) - k_1 - \mu_2\frac{p_s - p}{d}\right], \tag{6.113}$$

$$\frac{d\sigma_d}{dt} = \frac{1}{\nu_d}[k_1 - w(\sigma_d - \sigma_s)], \tag{6.114}$$

$$\frac{d\sigma_s}{dt} = \frac{1}{\nu_s}[w(\alpha_d - \alpha_s) - k_2], \tag{6.115}$$

$$\frac{d\alpha_d}{dt} = \frac{1}{\nu_d}[k_2 - w(\alpha_d - \alpha_s)], \tag{6.116}$$

$$p_s = 2k_4\frac{h_s^2}{\alpha_s - h_s}, \tag{6.117}$$

where,

$$h_s = \frac{\sigma_s - (\sigma_s^2 - k_3\alpha_s(2\sigma_s - \alpha_s))^{0.5}}{k_3}.$$

Let $d = 8.64$, $\mu_1 = 495$, $\mu_2 = 0.0495$, $\nu_s = 0.12$, $\nu_d = 1.23$, $w = 0.001$, $k_1 = 2.19x10^{-4}$, $k_2 = 6.12x10^{-5}$, $k_3 = 0.997148$, and $k_4 = 6.79x10^{-2}$.

Initially, $t = 0$, $p = 1.00$, $\sigma_s = 2.01$, $\sigma_d = 2.23$, $\alpha_s = 2.20$, and $\alpha_d = 2.26$. The function $f(t)$ is

$$f(t) = 2e^{-\frac{(t-1988)^2}{441}} + 10.2e^{-\frac{(t-2100)^2}{9216}} + 2.7e^{-\frac{(t-2265)^2}{3249}}. \tag{6.118}$$

Compare the predicted results using modified Euler method and RK4 method. Plot p, σ_s, and σ_d versus time.

Project 2: Coronavirus disease (COVID-19) is considered epidemic from 2019 and onward. Hence, an epidemiological model can be developed as

$$P = S(t) + F(t) + L(t) + I(t), \tag{6.119}$$

where P is the population size. $F(t)$, $S(t)$, $L(t)$, and $I(t)$ are the infected, susceptible, latent (infected but clinically unpronounced), and immune persons, respectively. The percentage of each category can be expressed as $y_1 = S(t)/P$, $y_2 = L(t)/P$, and $y_3 = F(t)/P$. Hence, $I(t)/P = 1 - y - 1 - y_2 - y_3$. The model equations are

$$\frac{dy_1}{dt} = \alpha - \beta(t)y_1 y_3, \tag{6.120}$$

$$\frac{dy_2}{dt} = \beta(t)y_1 y_3 - \gamma y_2, \tag{6.121}$$

$$\frac{dy_3}{dt} = \gamma y_2 - \delta y_3, \tag{6.122}$$

where $\beta(t) = 1580\,(1 + \cos(2\pi t))$. Let the fatality rate be $\alpha = 1.2 \times 10^{-2}$, $\mu = 0.2$, $\gamma = 50$, and $\delta = 100$.

Initial condition for the vector $Y = (y_1, y_2, y_3)$ is $(0.01, 0.02, 0.05)$. Compute and plot Y as function of time.

Project 3: The double pendulum problem, illustrated in Fig. 6.3, is described by the following system of ordinary differential equations:

$$\begin{cases} (m_1 + m_2)\, l_1 \ddot{\theta}_1 + m_2 l_2 \ddot{\theta}_2 \cos(\theta_1 - \theta_2) + m_2 l_2 \dot{\theta}_2^2 \sin(\theta_1 - \theta_2) \\ \quad + (m_1 + m_2)\, g \sin\theta_1 = 0, \\ l_1 \ddot{\theta}_1 \cos(\theta_1 - \theta_2) + l_2 \ddot{\theta}_2 - l_1 \dot{\theta}_1^2 \sin(\theta_1 - \theta_2) + g \sin\theta_2 = 0, \end{cases} \tag{6.123}$$

where the dot indicates the time derivative.

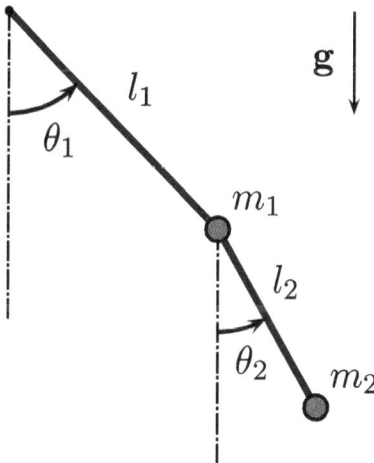

Figure 6.3: Double pendulum.

1. Get the explicit expression of $\ddot{\theta}_1$ and $\ddot{\theta}_2$, where the other parameters are assumed to be known. (*Hint*: Consider Eq. (6.123) as a two-by-two system and use Cramer's rule, see Section 2.4.)
2. Express the new shape of the problem as follows:

$$\frac{d\mathbf{Y}}{dt} = \boldsymbol{f}(\mathbf{Y}), \qquad (6.124)$$

where the components of vector $\mathbf{Y}$ are such that $(y_1, y_2, y_3, y_4) = (\theta_1, \dot{\theta}_1, \theta_2, \dot{\theta}_2)$.
3. Set $l_1 = 0.3\,\text{m}$, $l_2 = 0.2\,\text{m}$, $m_1 = 0.5\,\text{kg}$, $m_2 = 2\,\text{kg}$, $\theta_1(0) = \pi/4$, $\theta_2(0) = \pi/3$, $\dot{\theta}_1(0) = \dot{\theta}_2(0) = 0\,\text{s}^{-1}$, and $g = 9.81\,\text{ms}^{-2}$, then implement a program (or script) that solves the latter problem using modified Euler and RK4 methods up to time $t = 30\,\text{s}$. Test many time steps in order to get a non-diverging solution.

Extra Reading

J. Hadamard. Sur les problèmes aux dérivés partielles et leur signification physique. *Princeton University Bulletin*, 13:49–52, 1902.
P. D. Lax and R. D. Richtmyer. Survey of the stability of linear finite difference equations. *Communications on Pure and Applied Mathematics*, 9(2):267–293, 1956. doi: https://doi.org/10.1002/cpa.3160090206. URL https://onlinelibrary.wiley.com/doi/abs/10.1002/cpa.3160090206.
J. C. G. Walker. *Numerical Adventures with Geochemical Cycles*. Oxford University Press, 1st edition, 1990.

Chapter 7

Ordinary Differential Equations: Boundary Value Problems

7.1 Introduction

In this class of problems, the boundary conditions of the domain dictate the solution. In other words, information at the boundary instantly propagates into the domain with an infinite speed. For a one-dimensional domain, we have two boundaries. Therefore, the equation must be second order or higher. The method of solution can be somehow different from the method of solution of initial value problems. In general, there are two classes of methods of solution:

- shooting method;
- equilibrium methods, such as finite difference or volume methods.

In the shooting method, the problem is converted into an initial value problem with one (or more) missing boundary condition. The solution marches the other boundary by guessing the missing initial condition, expecting that such a guess leads to (more or less) the correct boundary condition on the other end. Interpolation or any of the methods explained in Chapter 4 will be used to improve the guess value and shooting repeatedly. If the problem is linear, then within exactly two guesses, the correct solution can be obtained. However, if the problem is nonlinear, the shooting is repeated until the ultimate solution hits the correct boundary condition within a prescribed tolerance.

In the following sections, the shooting and equilibrium methods are discussed in detail.

7.2   Linear Shooting Method

Let us consider the following linear boundary value problem:

$$
\begin{cases}
\dfrac{d^2 f}{dx^2} + P(x)\dfrac{df}{dx} + Q(x)f = R(x), \\[2mm]
f(a) = \alpha, \\[2mm]
f(b) = \beta,
\end{cases}
\tag{7.1}
$$

where $P(x)$, $Q(x)$, and $R(x)$ are continuous functions in $[a, b]$. In these conditions, it can be shown that the solution to this problem exists and is unique.

In order to solve problem (7.1), let us consider the following initial value problems instead:

$$
\begin{cases}
\dfrac{d^2 f_1}{dx^2} + P(x)\dfrac{df_1}{dx} + Q(x)f_1 = R(x), \\[2mm]
f_1(a) = \alpha, \\[2mm]
\dfrac{df_1}{dx}(a) = \beta_1
\end{cases}
\tag{7.2}
$$

and

$$
\begin{cases}
\dfrac{d^2 f_2}{dx^2} + P(x)\dfrac{df_2}{dx} + Q(x)f_1 = 0, \\[2mm]
f_2(a) = 0, \\[2mm]
\dfrac{df_2}{dx}(a) = \beta_2,
\end{cases}
\tag{7.3}
$$

which can be solved using any method among those presented in Chapter 6. Since the equation is linear, superposition principle is valid. Hence, by linear interpolation, the correct values of $f(b)$ can be found. Let us set $b_1 = f_1(b)$ and $b_2 = f_2(b)$. Hence, the correct initial derivative that leads to $f(b) = \beta$ is

$$\frac{df(a)}{dx} = \beta_1 + \frac{\beta - b_1}{b_2} b_1. \tag{7.4}$$

And the solution reads

$$f(x) = f_1(x) + \frac{\beta - b_1}{b_2} f_2(x). \tag{7.5}$$

7.3 Nonlinear Shooting Method

The most general shape of a second-order nonlinear boundary value problem can be cast as

$$\begin{cases} f'' = g\left(x, f, f'\right) & \text{for } a \leqslant x \leqslant b, \\ f(a) = \alpha, \\ f(b) = \beta, \end{cases} \tag{7.6}$$

where the dash symbol is for the differentiation with respect to x, that is $f' = \frac{df}{dx}$, $f'' = \frac{d^2 f}{dx^2}$, and g is a function with continuous first derivatives with respect to x, f, and f'. This notation holds throughout this section.

Let us introduce the parameter τ such that the problem (7.6) is recast as follows:

$$\begin{cases} \tilde{f}'' = g\left(x, \tilde{f}, \tilde{f}'\right) & \text{for } a \leqslant x \leqslant b, \\ \tilde{f}(a, \tau_k) = \alpha, \\ \tilde{f}'(a, \tau_k) = \tau_k, \end{cases} \tag{7.7}$$

where we have introduced a new function $\tilde{f} = \tilde{f}(x, \tau)$ and $(\tau_k)_{k \in \mathbb{N}}$ is a series we have to specify. The fundamental question is how to define the $(\tau_k)_{k \in \mathbb{N}}$ series in a way that its limit,

$$\lim_{k \to \infty} \tau_k = \bar{\tau}, \tag{7.8}$$

is such that

$$\tilde{f}(b, \bar{\tau}) = \beta \tag{7.9}$$

or equivalently

$$\lim_{k \to \infty} \tilde{f}(b, \tau_k) = \beta \tag{7.10}$$

and subsequently makes $\tilde{f}(x, \bar{\tau})$ the solution we are looking for? Actually, we can use Eq. (7.9) itself to define the series $(\tau_k)_{k \in \mathbb{N}}$ and so retrieve the recurrence relation. Here, we can regard k as the index of the kth correction of the solution.

7.3.1   Secant method

As previously mentioned, we intend to guess the value of $\tau = \tilde{f}'(a, \tau)$ and try to correct this value iteratively and bring function $\tilde{f}(x, \tau)$, at each iteration, even closer to the target solution of problem (7.6), that is, the solution of Eq. (7.9).

In order to improve the guess from one iteration to another (and so move forward with the series), we can use the secant method presented in Section 4.29. This simply reads

$$\tau_{k+1} = \tau_k - \omega \frac{\tilde{f}(b, \tau_k) - \beta}{\dfrac{\tilde{f}(b, \tau_k) - \tilde{f}(b, \tau_{k-1})}{\tau_k - \tau_{k-1}}}. \tag{7.11}$$

So, the method relies on providing **two** initial guesses, τ_0 and τ_1, then solving problem (7.7) for each guess. Afterward, τ_{k+1} is updated using (7.11), and the process is repeated until convergence, *viz.*, $\tau_{k+1} - \tau_k \leqslant \varepsilon$, or equivalently $f_{\tau_k}(b) - \beta \leqslant \varepsilon$, where ε is the tolerance set by the user. If the secant method converges, it converges linearly. The secant method can be made convergent or its convergence accelerated by considering under- or over-relaxation. Like the secant method applied to classical functions, if the method converges, it converges linearly.

7.3.2   Newton–Raphson method

Here, we intend to meliorate the shooting condition. An effective way to solve Eq. (7.9) is using Newton–Raphson method, presented in Section 4.3, instead. So, the series $(\tau_k)_{k \in \mathbb{N}}$ is simply given by the following recurrence relation:

$$\tau_{k+1} = \tau_k - \omega \frac{\tilde{f}(b, \tau_k) - \beta}{\dfrac{\partial \tilde{f}}{\partial \tau}(b, \tau_k)}, \tag{7.12}$$

where ω is a relaxation factor (optional). But a last difficulty remains. In (7.12), how to get the value of $\dfrac{\partial \tilde{f}}{\partial \tau}(b, \tau_k)$?

Differentiating the equations of problem (7.7) with respect to τ, assuming the order of differentiation to be reversible (this is called Young's theorem), rearranging, and defining the new function $\tilde{h}(x, \tau) = \frac{\partial \tilde{f}}{\partial \tau}$ yields the following **adjoined** problem:

$$\begin{cases} \tilde{h}'' = \dfrac{\partial g}{\partial \tilde{f}}\left(x, \tilde{f}, \tilde{f}'\right)\tilde{h} + \dfrac{\partial g}{\partial \tilde{f}'}\left(x, \tilde{f}, \tilde{f}'\right)\tilde{h}' & \text{for } a \leqslant x \leqslant b, \\ \tilde{h}(a, \tau_k) = 0, \\ \tilde{h}'(a, \tau_k) = 1. \end{cases} \qquad (7.13)$$

In that case, $\dfrac{\partial \tilde{f}}{\partial \tau}(b, \tau_k) = \tilde{h}(b, \tau_k)$, by virtue of the definition of function $\tilde{h}$ and the recurrence relation (7.12), becomes

$$\tau_{k+1} = \tau_k - \omega\frac{\tilde{f}(b, \tau_k) - \beta}{\tilde{h}(b, \tau_k)}. \qquad (7.14)$$

Although partial derivatives are involved in problem (7.13), one must keep in mind that: (i) no partial derivatives with respect to τ appear in problem (7.13) and (ii) the problem (7.7) needs to be solved at each iteration at a given and fixed value of τ, τ_k. So, it still remains an ordinary differential problem with one independent variable, namely x. Considering the problem this way is called **the method of lines**.

In summary, the method relies on providing an initial guess of τ_0, then solving at the same time problems (7.7) and (7.13), update τ_{k+1} using (7.14) that the solution of problem (7.13) provides and repeat until $\tau_{k+1} - \tau_k \leqslant \varepsilon$, where ε is the tolerance fixed by the user.

The approach seems to be a real *tour de force* in order to solve nonlinear boundary value problems using marching-like methods. But the method is so elegant and powerful that it deserves to be mentioned and, whenever possible, used. The main usefulness of such a method is its ability to refine the step size on the fly, i.e., during the computation, according to the behavior of the calculated solution (steep slope, for instance), which can be a desired property when solving stiff and boundary-layer problems. Another feature of this method (see Section 4.3) is that it converges quadratically and so much faster than the secant method.

7.4   Finite Difference Method

7.4.1   Planar problems

The domain of solution has to be divided into (say n) segments or assigned nodes at equal or unequal intervals. A general second-order ODE can be written as

$$\frac{d^2 f}{dx^2} + P\frac{df}{dx} + Qf = F(x, f), \tag{7.15}$$

where P and Q can be functions of x and f (but not of their derivatives). For simplicity, we assume that P and Q are functions of x only. To close the problem, two boundary conditions need to be specified: At $x = 0$, $f = f_0$ and at $x = l$, $f = f_l$, for example. The other kinds of boundary conditions will be discussed later on.

Using second-order finite difference and for constant Δx, the ODE converts into a system of algebraic equations:

$$\frac{f_{i+1} - 2f_i + f_{i-1}}{\Delta x^2} + P(x_i)\frac{f_{i+1} - f_{i-1}}{2\Delta x} + Q(x_i)f_i = F(x_i, f_i). \tag{7.16}$$

Rearranging the equation,

$$\left(1 - \frac{P(x_i)\Delta x}{2}\right) f_{i-1} + \left(-2 + Q_i\Delta x^2\right) f_i$$

$$+ \left(1 + \frac{P(x_i)\Delta x}{2}\right) f_{i+1} = F(x_i, f_i)\Delta x^2. \tag{7.17}$$

Such a discretization leads to solving tri-diagonal matrices using Thomas algorithm, for instance, see Section 2.4.2.

7.4.2   Axisymmetric problems

A few engineering and scientific problems involve cylindrical and spherical geometries, such as heat conduction in pipes and nuclear fuel rods. A one-dimensional axisymmetric problem in cylindrical coordinate system can be expressed as

$$\frac{1}{r}\frac{d}{dr}\left(r\frac{df}{dr}\right) = S. \tag{7.18}$$

The left-hand term can be expanded as

$$\frac{1}{r}\frac{d}{dr}\left(r\frac{df}{dr}\right) = \frac{d^2 f}{dr^2} + \frac{1}{r}\frac{df}{dr}. \tag{7.19}$$

Using central difference method to approximate the above differential equation yields

$$\frac{f_{i+1} - 2f_i + f_{i-1}}{\Delta r^2} + \frac{f_{i+1} - f_{i-1}}{r_i \Delta r} = S_i. \tag{7.20}$$

The symmetry condition at $r = 0$ imposes $\frac{df}{dr} = 0$. The condition at $r = 0$ can be approximated as $\frac{f_{i+1} - f_i}{\Delta r}$; however, the approximation is first order, which is not recommended. Alternatively, the governing equation can be used at $r = 0$. The problem of the term $\frac{1}{r}\frac{df}{dr}$ can be resolved by using L'Hospital's rule of limit:

$$\lim_{r \to 0} \frac{1}{r}\frac{df}{dr} = \lim_{r \to 0} \frac{\frac{d}{dr}\left(\frac{df}{dr}\right)}{\frac{d}{dr}(r)} = \frac{d^2 f}{dr^2}\bigg|_{r=0}. \tag{7.21}$$

Hence, the governing equation at $r = 0$ will be

$$2\frac{d^2 f}{dr^2} = S. \tag{7.22}$$

Using central difference approximation,

$$2\frac{f_{i+1} - 2f_i + f_{i-1}}{\Delta r^2} = S_i. \tag{7.23}$$

Note that due to symmetry, $f_{i-1} = f_{i+1}$. So, we retrieve the following equation:

$$2\frac{2f_{i+1} - 2f_i}{\Delta r^2} = S_i. \tag{7.24}$$

7.5   Boundary Conditions

Since the mathematical models of boundary value problems yield differentials with second-order derivatives, two boundary conditions are needed to close the solution. The following are three kinds of boundary conditions:

1. The function value is specified at the boundary (Dirichlet boundary condition). This is the simplest boundary condition, no special treatment is needed.
2. The first derivative of the function is given (Neumann boundary condition). There are a few options to deal with this kind of boundary condition. One option is using first-order approximation of the derivative. For example, the derivative is given on the left-hand boundary, see Fig. 7.1; the derivative can be approximated using finite difference as

$$\frac{df}{dx} = \frac{f_1 - f_0}{\Delta x}. \tag{7.25}$$

However, using first-order boundary condition will reduce the overall accuracy of the solution to first order. Hence, it is not recommended. Second-order approximation is recommended. One way to approximate the first derivative is by using extra internal nodes (node 2, beside nodes 0 and 1). Or simply, we can artificially extend the domain by one ghost node outside of the domain, as shown in Fig. 7.1. Second-order finite difference can be used to approximate the derivative as

$$\frac{df}{dx} = \frac{f_1 - f_{-1}}{2\Delta x}. \tag{7.26}$$

The approximation provides an equation for f_{-1}.
3. A boundary that combines the values of function and its derivative (Cauchy condition or Robin condition). This kind of boundary conditions mainly arises from balance equations at the boundary or at the interface. For instance, the balance between a heat conduction and heat convection at the boundary yields

$$-k\frac{dT}{dn} = h(T - T_a), \tag{7.27}$$

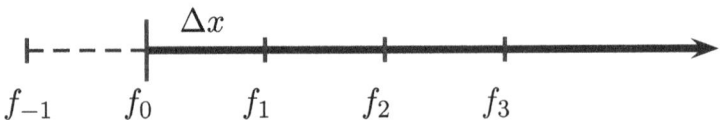

Figure 7.1:   Applying boundary condition by introducing a ghost node.

where k, T, n, h, and T_a are thermal conductivity, temperature, direction normal to the boundary, heat transfer coefficient, and ambient temperature, respectively. Regular finite difference can be used to approximate the first derivative.

Remark: In general, second-order accurate scheme is sufficient to obtain reliable results. Of course, we can use higher-order schemes without any problem, except the one that arises at the boundary. However, it is better to use smaller h (Δx) values rather than using higher-order schemes beyond, say, second-order accurate schemes.

7.5.1 Modified equation

Let us solve the following equation by using second-order accurate scheme:

$$\frac{d^2 f}{dx^2} + a\frac{df}{dx} + bx = c, \qquad (7.28)$$

where a, b, and c are constants. The finite difference approximation yields

$$\frac{f_{i+1} - 2f_i + f_{i-1}}{h^2} + a\frac{f_{i+1} - f_{i-1}}{2h} + bx_i = c. \qquad (7.29)$$

Rearranging gives

$$f_i = \frac{h^2}{2}\left[\left(\frac{1}{h^2} + \frac{a}{2h}\right)f_{i+1} + \left(\frac{1}{h^2} - \frac{a}{2h}\right)f_{i-1} + bx_i - c\right]. \qquad (7.30)$$

Expanding f_{i+1} and f_{i-1}, $f_{i+1} = f_i + f'h + f''h^2/2 + f'''h^3/6 + \cdots$ and $f_{i-1} = f_i - f'h + f''h^2/2 - f'''h^3/6 + \cdots$ and substituting in Eq. (7.28) yields

$$f'' + af' + bx = c + ah^2 f'''/6. \qquad (7.31)$$

Comparing with the original equation, there is an extra term, $ah^2 f'''/6$. However, as h approaches very small value, the extra term becomes insignificant. Though dispersive, the scheme is consistent.

### 7.5.2	Example

Solve the following boundary value problem:

$$a\frac{df}{dx} = b\frac{d^2 f}{dx^2}. \tag{7.32}$$

Boundary conditions: $f(0) = 0$ and $f(L) = 1$.

Use second-order finite difference method. Set a, b, and L to be equal to 0.2, 0.6, and 2, respectively. Present the results for $\Delta x = 0.4$ and 0.2. Compare the prediction with the analytical solution. Note that the solution will diverge on using $\Delta x = 0.4$. The remedy is to use a forward difference for the first derivative.

## 7.6	Finite Volume Method

Finite volume method (FVM) is another technique to solve boundary value problems. The concept of finite volume or control volume approach is not new for engineering and science students. For instance, in analyzing a system, from thermodynamics or fluid mechanics standpoint, a control volume is drawn around the system to distinguish it from the environment. The system performance can be judged from inlet and outlet, without knowledge of the details of the system. The principle of FVM is similar to where control volume approach is substituted by many finite volumes rather than a unique and global volume. The domain is divided into (not necessarily uniform) control volumes called cells, see Fig. 7.2. The procedure of setting the control volume is as follows. First, divide the domain into control volumes (dashed lines). Then, set the representative of the

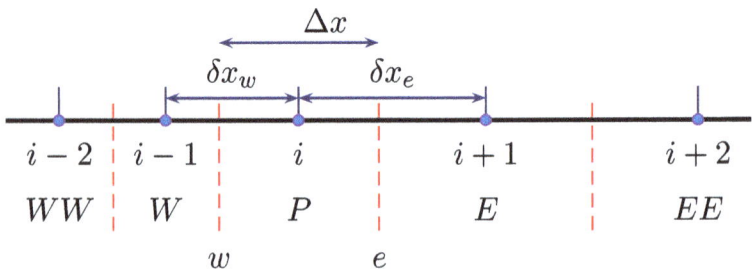

Figure 7.2:	Non-uniform FVM arrangement: local and global notations.

control volume values at the middle of the control volumes (dots). In the control volume approach, we assume that the values of the dependent variables are constant within the control volume, which is represented by the central point (dot).

It is possible to use notations such as $i-1$, i, and $i+1$ for the west, central, and east cells. However, it is more convent to use west, W, central, P, and east, E, for illustration purposes. We call this local notation. We can also use notations w and e to refer for the west and east interfaces (boundaries) of a given control volume, respectively.

There are two approaches in using FVM, either using the physics of the problem or using the governing equations. In the following sections, examples will be given to illustrate the mentioned approaches.

7.6.1   Physics approach

Let us assume that the temperature distribution along a pipe needs to be evaluated. The flow enters the pipe of diameter D with a flow rate of $\dot{m}$ ($= \rho v A$, where ρ is the fluid density, v the flow velocity, and A the cross-sectional area of the pipe) and at a temperature of T_{in}. The pipe length is L. An electric heater surrounding the pipe supplies total heat flux of q_{elec} (in W, and so the uniform heat flux density of it is $q''_{\text{elec}} = \frac{q_{\text{elec}}}{\pi D L}$ W/m^2. The question is, what is the outlet temperature T_{out} of the fluid?

First, we assume that the inlet temperature T_{in} is known and that the temperature normal gradient at the outlet is zero, that is $\frac{\partial T}{\partial x} = 0$, we use linear extrapolation.

Next, we divide the pipe into n small segments, each say Δx long. Hence, the number of segments is such that $\Delta x = L/n$, see Fig. 7.3. The energy equation will be applied to each segment. The first law of thermodynamics states that power in, minus power out, plus generated power equals exactly the rate of change of energy of the system. (Power here refers to energy per unit time.)

So, the balance states that

total power into the system	$-$	total power out of the system	$+$	power generated within the system	$=$	rate of change of energy of the system,

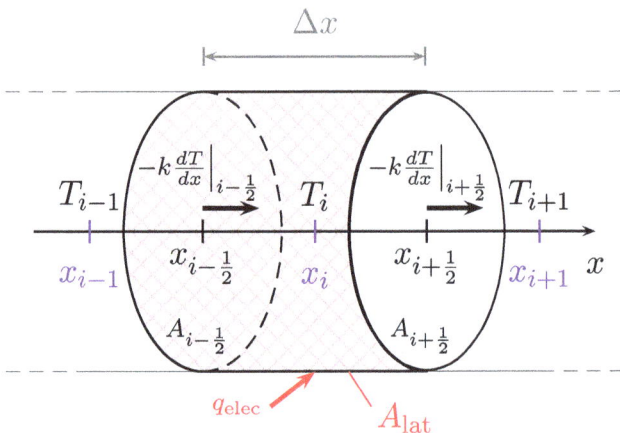

Figure 7.3:   Sketch of one-dimensional finite volume cells.

i.e.,

$$\sum_{\text{in}} \dot{E} - \sum_{\text{out}} \dot{E} + \sum_{\text{gen}} \dot{Q} = \frac{dE}{dt}\bigg|_{\text{system}}, \qquad (7.33)$$

where $\dot{E}$ is the rate of energy (power) in units of J/s or Watt. Under steady state, the right-hand term cancels. For the given problem, there is no heat generation within the system. For the steady-state case, the energy into the system minus energy out of the system must be equal to zero:

$$\left(\rho v c_p T_{i-\frac{1}{2}} - k \frac{dT}{dx}\bigg|_{x_{i-\frac{1}{2}}} \right) A_{i-\frac{1}{2}} + q''_{\text{elec}} A_{\text{lat}}$$

$$- \left(\rho v c_p T_{i+\frac{1}{2}} - k \frac{dT}{dx}\bigg|_{x_{i+\frac{1}{2}}} \right) A_{i+\frac{1}{2}} = 0. \qquad (7.34)$$

Note that Fourier law for heat conduction states that $\dot{E} = -kA\frac{dT}{dx}$. The heat flux density q''_{elec} is $\frac{q_{\text{elec}}}{\pi D L}$. Also, note that $i-1/2$ and $i+1/2$ represent the west and east boundaries of the control volume, respectively.

By taking the mean temperature for the advective term (which may not always be a good idea, for stability issues) and using a second-order approximation of the derivative $\frac{dT}{dx}$ for the conductive term, we get

$$\left(\rho c_p v \frac{T_{i-1} + T_i}{2} - k\frac{T_i - T_{i-1}}{\Delta x}\right)\frac{\pi D^2}{4}$$

$$- \left(\rho v c_p \frac{T_i + T_{i+1}}{2} - k\frac{T_{i+1} - T_i}{\Delta x}\right)\frac{\pi D^2}{4} = -q''_{\text{elec}}\pi D\Delta x, \quad (7.35)$$

which rearranged yields

$$\left(\frac{\rho c_p v}{2} + \frac{k}{\Delta x}\right)T_{i-1} - \frac{2k}{\Delta x}T_i + \left(-\frac{\rho v c_p}{2} + \frac{k}{\Delta x}\right)T_{i+1} = -\frac{4q_{\text{elec}}\Delta x}{D}.$$

$$(7.36)$$

This equation holds for any cell i, $1 < i < n$. For the first cell (that is $i = 1$), the west cell does not exist. But the temperature at the west face of that cell is defined by the boundary condition $T = T_{\text{in}}$. So, the temperature $T_{i-\frac{1}{2}}$ appearing in the advection term in Eq. (7.34) is replaced by this known temperature T_{in}. Similarly, the temperature gradient appearing in the conduction term of the same Eq. (7.34) is replaced by a first-order approximation using the west face temperature T_{in} and cell temperature T_1. This yields the following equation (where i is replaced by 1):

$$\left(\rho c_p v T_{\text{in}} - k\frac{T_1 - T_{\text{in}}}{2\Delta x}\right)\frac{\pi D^2}{4}$$

$$- \left(\rho v c_p \frac{T_1 + T_2}{2} - k\frac{T_2 - T_1}{\Delta x}\right)\frac{\pi D^2}{4} = -q''_{\text{elec}}\pi D\Delta x, \quad (7.37)$$

which rearranged reads

$$\left(-\frac{\rho v c_p}{2} - \frac{3k}{2\Delta x}\right)T_1 + \left(-\frac{\rho v c_p}{2} + \frac{k}{\Delta x}\right)T_2$$

$$= -\frac{4q''_{\text{elec}}\Delta x}{D} - \left(\rho v c_p + \frac{k}{2\Delta x}\right)\frac{4T\text{in}}{\pi D^2}. \quad (7.38)$$

In an analogous manner, for the last cell (that is $i = n$), the east cell does not exist. But the temperature gradient at the east face of that cell is defined by the boundary condition $\frac{dT}{dx} = 0$. So, the

temperature gradient $\frac{dT}{dx}\big|_{x_{i+\frac{1}{2}}}$ appearing in the conduction term in
Eq. (7.34) is replaced by this known temperature gradient of 0. Now,
if we apply once more the boundary condition and approach it using
a first-order approximation, we get $\frac{T_{out}-T_n}{2\Delta x} = 0$, that is $T_{out} = T_n$. So,
we can replace the east face temperature appearing in the advection
term of cell n simply by T_n. This yields the following equation (where
i was replaced by n):

$$\left(\rho c_p v \frac{T_{n-1}+T_n}{2} - k\frac{T_n-T_{n-1}}{\Delta x}\right)\frac{\pi D^2}{4}$$

$$- (\rho v c_p T_n)\frac{\pi D^2}{4} = -q''_{elec}\pi D\Delta x, \tag{7.39}$$

which rearranged reads

$$\left(\frac{\rho c_p v}{2} + \frac{k}{\Delta x}\right)T_{n-1} + \left(-\frac{\rho c_p v}{2} - \frac{k}{\Delta x}\right)T_n = -\frac{4q''_{elec}\Delta x}{D}. \tag{7.40}$$

The set of Eqs. (7.36), $i = 2,\ldots,n-1$, (7.38), and (7.40) being a
tri-diagonal linear system of n equations with n unknowns, can be
solved by using Thomas algorithm described in Section 2.4.2. Once
solved, the temperature T_{out} is retrieved by setting $T_{out} = T_n$, which
answers the question asked earlier.

Important: Note that it is well known that taking the face temperature, appearing in the advection term, to be equal to the mean
temperature of its neighboring cells poses stability issues. But as
far as advection-conduction problems are concerned, such issues can
always be circumvented by reducing the space step Δx. To avoid the
possible stability issue, the advection term should reflect the physics.
Since the flow is entering from the left, the advection temperature
should be T_{i-1} and not the average $\frac{T_i+T_{i-1}}{2}$. (At least, more weight
can be given to upstream temperature, T_{i-1}.) For the flow leaving
the control volume, the temperature at the east interface should be
approximated by T_i and not value, $\frac{T_i+T_{i+1}}{2}$. However, the order of
accuracy of the solution becomes first order. To increase order of
accuracy, it is recommended to use more nodes, with schemes biased
toward the flow direction.

7.6.2 Local governing equation approach

The physics of the problem is already modeled and the local governing equation is given with a set of boundary and initial conditions. For example, the heat conduction in a one-dimensional problem along a solid bar exposed to electric power is given by the following differential equation:

$$\rho c_p v \frac{dT}{dx} = \frac{d}{dx}\left(k\frac{dT}{dx}\right) + q'''_{elec}, \qquad (7.41)$$

where q'''_{elec} is the equivalent heat generation per unit volume, which is equal to $q_{elec}/(AL)$. For the circular tube, $A = \pi D^2/4$. Note that in the above equation, we did not assume that the thermal conductivity is a constant.

To use finite volume method, the above equation is multiplied by volume $\frac{\pi D^2}{4}dx$ since $\frac{\pi D^2}{4}$ is constant, which cancels. Consequently, we use dx and integrate, which reads

$$\int_w^e \rho c_p v \frac{dT}{dx} dx = \int_w^e \frac{d}{dx}(k\frac{dT}{dx})dx + \int_w^e \frac{q_{elec}}{AL}dx. \qquad (7.42)$$

By performing the integration, the above equation yields

$$\rho c_p v T|_e - \rho c_p v T|_w = k\left.\frac{dT}{dx}\right|_e - k\left.\frac{dT}{dx}\right|_w + \frac{q_{elec}\Delta x}{AL}, \qquad (7.43)$$

which is exactly equivalent to Eq. (7.34). Yet, we did not approximate any terms. However, we do not have the value of temperature at the interfaces, w and e. Hence, we have to approximate the value of T and gradient of T (flux) at the west and east interfaces using the temperature at the cells' center. For uniform cells, δxw and δxe are equal to Δx. Therefore,

$$T_e = \frac{T_E + T_p}{2} \quad \text{and} \quad \left.\frac{dT}{dx}\right|_e = \frac{T_E - T_P}{\Delta x}. \qquad (7.44)$$

In i-indices notation,

$$T_{i+1/2} = \frac{T_{i+1} + T_i}{2} \quad \text{and} \quad \left.\frac{dT}{dx}\right|_{i+1/2} = \frac{T_{i+1} - T_i}{\Delta x}. \qquad (7.45)$$

Similarly, for the interface w $(i - 1/2)$.

For non-uniform cells, linear interpolation can be used, i.e.,

$$T_e = \frac{T_E \Delta x/2 + T_P(\delta xe - \Delta x/2)}{\delta xe}. \tag{7.46}$$

Such an approach ensures second-order accuracy. However, it causes stability problem for relatively high velocities, v. To remedy the stability problem, we need to bias the scheme toward the flow direction. For example, if the flow is from left to right (v is positive), T_w should be replaced by T_W and T_e by T_P. For flow from right to left (v is negative), T_w should be replaced by T_P and T_e by T_E.

The slope at the east interface can be approximated as

$$\left.\frac{dT}{dx}\right|_e = \frac{T_E - T_P}{\delta xe}. \tag{7.47}$$

Similarly, the slope at the west interface can be approximated as

$$\left.\frac{dT}{dx}\right|_w = \frac{T_P - T_W}{\delta xw}. \tag{7.48}$$

7.6.3   Example: Finite difference and shooting method

The Crocco–Wang (singular) problem is a more amenable form of the Blasius equation describing the boundary layer developing over a flat plate. Along with the associated boundary conditions, this problem reads

$$\begin{cases} y'' + \gamma \dfrac{x}{y} = 0 & 0 \leqslant x \leqslant 1, \\ y'(0) = 0, \\ y(1) = 0, \end{cases} \tag{7.49}$$

where γ is a strictly positive real number.

We intend to apply here and compare the finite difference and the nonlinear shooting methods. Using the first- and second-order finite difference schemes, the discretized equations read

$$\begin{cases} \dfrac{2y_2 - 2y_1}{2\Delta x} = 0 & i = 1, \\ \dfrac{y_{i-1} - 2y_i + y_{i+1}}{\Delta x^2} + \gamma \dfrac{x_i}{y_i} = 0 & i = 2, \ldots, N-1, \\ y_N = 0, \end{cases} \tag{7.50}$$

where N is the number of discretization points and Δx is the step size given by $\Delta x = \frac{1}{N-1}$. Because of its nonlinearity, the problem has to be solved using Newton–Raphson method.

The problem can also be solved using the nonlinear shooting method, which in turn solves, in this particular case, the following problem:

$$
\begin{cases}
y'' = -\dfrac{\gamma x}{y}, \\[2mm]
h'' = \dfrac{\gamma x h}{y^2}, \\[2mm]
y(1) = 0, \\
y'(1) = \tau, \\
h(1) = 0, \\
h'(1) = 1,
\end{cases}
\qquad (7.51)
$$

where τ is updated until convergence using the following relationship:

$$
\tau_{k+1} = \tau_k - \omega \frac{y(1)}{h(1)}. \qquad (7.52)
$$

The initial value problem (7.51) can be solved using Runge–Kutta method, for instance.

Figure 7.4 shows the solution obtained using FDM and shooting method for different values of parameter γ, and Table 7.1 shows the value of $y(0)$ predicted by both methods compared to benchmark solution given by Cortell. In the computation, Δx and ω are set to $\frac{1}{200}$ and 0.5, respectively.

7.6.4 Example: Finite volume method

An electric heater is used to heat the water at temperature 15°C by passing the water through a pipe of 5 cm in diameter with a flow rate of 0.01 kg/s. The pipe length is 1.2 m. If the heater operates at 1500 W, estimate the outlet temperature using finite volume method. Plot the temperature distribution along the pipe. Compare the predicted results with analytical solution. Use uniform control volume (cells), each 20 cm in length.

Solution: Figure 7.5 shows the numbering of the control volumes, where $\Delta x = 0.2$ m.

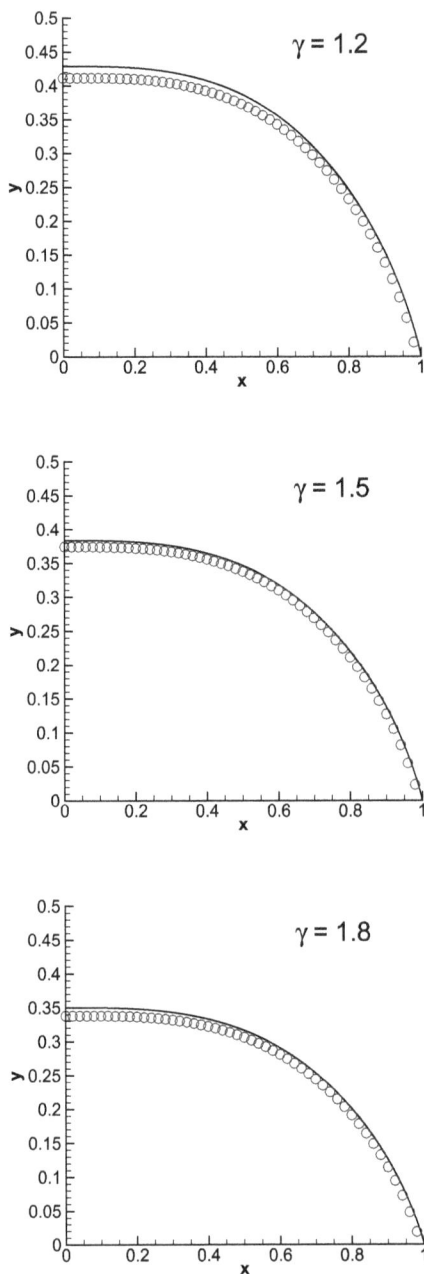

Figure 7.4:   Boundary layer problem solved using Crocco–Wang transformation. Comparison between the solutions given by FDM (circles) and shooting method (solid line) for different values of γ.

Table 7.1: Value of $y(0)$ for different values of the inverse of γ.

γ^{-1}	Shooting method	FDM	Reference (Cortell)
1.2	0.42874792	0.41176191	0.42868
1.5	0.38348380	0.37421601	0.38342
1.8	0.35007104	0.33810584	0.35002

Figure 7.5: FV arrangement for the heater problem.

We have seven unknowns (points 2, 3, 4, 5, 6, 7, and 8). In general, the advection term $(\dot{m}c_p)$ is higher than the conduction term. This also implies that the temperature at the entrance to the control volume is dictated by the upstream temperature. We can neglect the conduction effect. Hence, the energy balance will be

$$\dot{m}c_pT|_{\text{in}} - \dot{m}c_pT|_{\text{out}} + Q|_{\text{in}} = 0. \qquad (7.53)$$

Since $\dot{m}$ and c_p ($c_p = 4186$ J/kg.K) are assumed constant, the above equation can be rewritten for each control volume as

$$T|_{\text{in}} - T|_{\text{out}} + \frac{Q_{\text{in}}\Delta x}{\dot{m}c_pL} = 0 \qquad (7.54)$$

or

$$T_{\text{out}} = T_{\text{in}} + \frac{Q_{\text{in}}\Delta x}{\dot{m}c_pL}. \qquad (7.55)$$

Let us deal with each control volume:

CV 2: $T_2 = T_1 + 5.97 = 20.97$,
CV 3: $T_3 = T_2 + 5.97 = 26.94$,
CV 4: $T_4 = T3 + 5.97 = 32.91$,
CV 5: $T_5 = T_4 + 5.97 = 38.88$,
CV 6: $T_6 = T_5 + 5.97 = 44.85$,
CV 7: $T_7 = T6 + 5.97 = 50.82$,
Outlet: $T_8 = T_7 = 50.82$.

The analytical solution for $\dot{m}c_p \frac{dT}{dx} = Q/L$ is $T = T_{\text{in}} + \frac{Q}{L\dot{m}c_p}x$, which is

$$T = T_{\text{in}} + 29.86x \quad \text{for } x = 1.2, \ T = 50.83.$$

Note that neglecting the conduction term, the problem becomes an initial value problem. The question is, when we can neglect the conduction term and when we cannot?

As we mentioned before, it is better to non-dimensionalize the problem before solving it. Let us work the problem properly.

The governing equation is

$$\rho c_p V \frac{dT}{dx} = k \frac{d^2 T}{dx^2} + Q'''. \tag{7.56}$$

The boundary conditions are at $x = 0$, $T = T_i$ and at $x = L$, $\frac{dT}{dx} = 0$.

Here ρ and V are density, velocity and cross-sectional area of the pipe, respectively. Let $x^* = x/L$. Dividing the above equation by $\rho c_p V / L$ yields

$$\frac{dT}{dx^*} = \frac{k}{\rho c_p V L} \frac{d^2 T}{dx^{*2}} + \frac{LQ'''}{\rho c_p V}. \tag{7.57}$$

The thermal diffusivity is $\alpha = \frac{k}{\rho c_p}$. Note that the last term, $\frac{LQ'''}{\rho c_p V}$, is homogeneous in temperature. Hence, let $\theta = \frac{(T - T_i)\rho c_p V}{LQ'''}$. Also, multiply and divide the first term on the right-hand side by the kinematic viscosity of the fluid, ν:

$$\frac{d\theta}{dx^*} = \frac{1}{\text{Re}\,\text{Pr}} \frac{d^2\theta}{dx^{*2}} + 1. \tag{7.58}$$

Boundary conditions: $x^* = 0$, $\theta = 0$ and $x^* = 1$, $\frac{d\theta}{dx^*} = 0$. Here, $\text{Re} = \frac{VL}{\nu}$ is Reynolds number and $\text{Pr} = \frac{\nu}{\alpha}$ is Prandtl number, and the product of Re and Pr is called Peclet number, Pe. For the given water flow ($\alpha = 0.146 \times 10^{-6} \text{m}^2/\text{s}$), Pe $= 82.2 \times 10^3$. Now, it is clear that if Pe $\gg 1$, we can neglect the conduction term without introducing significant error to the solution. Also, the solution of the above equation is very general and can be applied to any length of the pipe and/or flow rate provided that Pe is the same.

The above equation can be solved by FVM as

$$\int_w^e \frac{d\theta}{dx^*}dx^* = \frac{1}{\text{Pe}}\int_w^e \frac{d^2\theta}{dx^{*2}}dx^* + \int_w^e 1 dx^*. \tag{7.59}$$

Performing the integration:

$$\theta_e - \theta_w = \frac{1}{\text{Pe}}\left(\frac{d\theta}{dx^*}\Big|_e - \frac{d\theta}{dx^*}\Big|_w\right) + \Delta x^*. \tag{7.60}$$

Like previously, we neither know the values of the dependent variable, θ, nor the derivative of θ at the interfaces of the control volume. Therefore, we need to approximate those unknowns. The physics of the problem is that the information flows toward the direction of the stream. Hence, if the flow is from the left to the right, then it is a good idea to approximate θ_e and θ_w by θ_P and θ_W, respectively. Again, this assumption is valid for Pe > 1. The derivatives can be approximated as

$$\frac{d\theta}{dx^*}\Big|_e = \frac{\theta_E - \theta_P}{\delta xe^*} \quad \text{and} \quad \frac{d\theta}{dx^*}\Big|_w = \frac{\theta_P - \theta_W}{\delta xw^*}. \tag{7.61}$$

Hence,

$$\theta_P - \theta_W = \frac{1}{\text{Pe}}\left(\frac{\theta_E - \theta_P}{\delta xe^*} - \frac{\theta_P - \theta_W}{\delta xw^*}\right) + \Delta x^*. \tag{7.62}$$

For uniform control volumes: $\Delta x^* = \delta xe^* = \delta xw^*$. Therefore,

$$\theta_P(1 + \frac{2}{\text{Pe}\Delta x^*}) = \frac{1}{\text{Pe}\Delta x^*}(\theta_E + \theta_W) + (\theta_W + \Delta x^*) \tag{7.63}$$

or

$$a_P\theta_P = a_E\theta_E + a_W\theta_W + S, \tag{7.64}$$

where $a_E = \frac{1}{\text{Pe}\Delta x^*} = 7.3 \times 10^{-5}$, $a_W = \frac{1}{\text{Pe}\Delta x^*}$, $a_P = a_E + a_W$, and $S = \theta_W + \Delta x^*$.

Note the following: a_P is the sum of a_E and a_P. Also, the coefficients of $a's$ are positive.

The derivatives at the boundary need to be treated differently because the distance between the boundary and the center of the control volume is half of Δx^*. For example, at the left boundary,

$$\frac{d\theta}{dx^*}\Big|_w = \frac{\theta_P - \theta_{\text{boundary}}}{0.5\Delta x^*}. \tag{7.65}$$

7.7   Problems

Problem 1: The equation that the deflection of an elastic rope satisfies reads

$$\begin{cases} P\,y'' = q & \text{for } 0 \leqslant x < L, \\ y = \alpha & \text{for } x = 0, \\ y = \beta & \text{for } x = L, \end{cases}$$

where $P > 0$ is the rope pre-stress and $q > 0$ the linear density of the rope.

Solve this equation using the linear shooting method. Let $L = 2.5$, $P = 2$, $q = 1.5$, $\alpha = 0$, and $\beta = 1.0$. Compare the numerical predictions with the analytical solution.

Problem 2: Inspired by the method presented in Section 7.2 to solve a linear boundary value problem, suggest a similar shooting method to solve the following problem:

$$\begin{cases} \dfrac{d^2 f}{dx^2} + P(x)\dfrac{df}{dx} + Q(x)f = R(x), \\ f(a) = \alpha, \\ \dfrac{df}{dx}(b) = \beta. \end{cases}$$

Implement and test this solution by setting $a = 0$, $b = 1$, $\alpha = 1$, $\beta = 2$, $P(x) = 1 - x^2$, $Q(x) = 2$, and $R(x) = x$.

Problem 3: Falkner–Skan boundary layer problem,

$$\begin{cases} f''' + f\,f'' + \beta(1 - f'^2) = 0 & \text{for } 0 \leqslant x < \infty, \\ f = f' = 0 & \text{for } x = 0, \\ f' = 1 & \text{for } x \to \infty, \end{cases}$$

describes the incompressible flow about a wedge of angle $\beta\pi$. (Dashes represent successive differentiation with respect to x.) The dimensionless velocity component along the wedge is given by $f'(x)$. This problem reduces to Blasius's boundary layer on setting $\beta = 0$.

Solve this problem using nonlinear shooting method and the finite difference and volume methods and compare the performance of both methods. (Set $\beta = \frac{2}{3}$.)

Problem 4: The same as Problem 3 with the compressible Falkner–Skan boundary layer problem:

$$\begin{cases} (\rho\,\mu\,f'')' + f\,f'' + \beta(h - f'^2) = 0 & \text{for } 0 \leqslant x < \infty, \\ (\rho\,\mu\,h')' + \Pr f\,h' = 0 & \text{for } 0 \leqslant x < \infty, \\ f = f' = 0 & \text{for } x =, 0 \\ f' = 1 & \text{for } x \to \infty, \\ h' = 1 & \text{for } x \to \infty, \end{cases}$$

with Pr the Prandlt number, $\rho = h^{-1}$, and $\mu = h^{2/3}$. The heat capacity is assumed constant in this model. (Set $\Pr = 0.7$ and $\beta = \frac{4}{3}$.)

Problem 5: The same as Problem 3 with Lane–Emden equation describing the density distribution of a self-gravitating, spherical symmetric polytropic fluid given by

$$\begin{cases} (r^2\theta')' + r^2\theta^n = 0 & \text{for } 0 \leqslant r < L, \\ \theta = \alpha & \text{for } x = 0, \\ \theta = \beta & \text{for } x = 0, \end{cases}$$

where r is the radius, θ a quantity related to the density, and $n \geq 1$ the polytropic index of the fluid. (Set $n = 2$, $L = 1$, $\alpha = 1$, and $\beta = \frac{3}{2}$.)

Problem 6: Counter-flow heat exchangers are used in many industrial applications. Figure 7.6 shows a simple plate heat exchanger. The product of the flow rate and the specific heat is given for both fluids, $\dot{m}_1 cp_1 = 300$ W/K and $\dot{m}_2 cp_2 = 500$ W/K. The heat exchanger length is 1.1 m. The hot (side 1) and cold (side 2) fluids entrance temperatures are 100 °C and 20 °C, respectively. Neglect the thickness of the separating plate. Assume that the heat exchange coefficients for the cold and hot fluids sides are 1000 W/m^{2}K and 200 W/m^{2}K, respectively. Predict the temperature distribution along the heat exchanger for both fluids.

Problem 7: The deflection in a thin rode is governed by the following equation:

$$\frac{d^2y}{dx^2} + \frac{F}{EI}\left[1 + \left(\frac{dy}{dx}\right)^2\right]^{1.5} y = 0, \qquad (7.66)$$

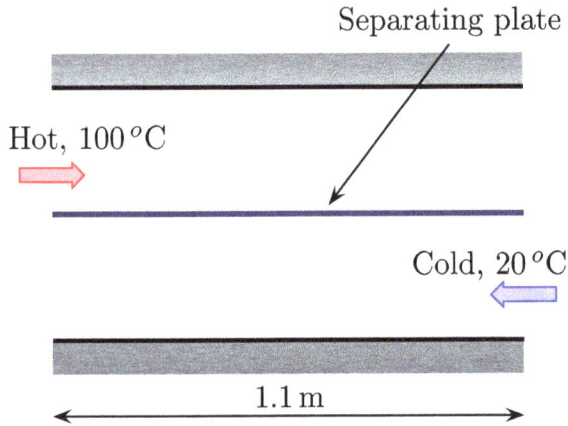

Figure 7.6:   A simple heat exchanger.

where E and I are Young's modulus and moment of inertia, respectively. The applied force is F. Let $E = 10^6$, $I = 0.02$, and $F = 100$. The deflection at both ends (boundary conditions), $x = 0$ and $x = 2$, is $y = 0$. Predict the deflection distribution along the rode.

Extra Reading

R. Cortell. Numerical solutions of the classical Blasius flat-plate problem. *Applied Mathematics and Computation*, 170(1):706–710, 2005. doi: https://doi.org/10.1016/j.amc.2004.12.037. URL https://www.sciencedirect.com/science/article/pii/S0096300305000111.

K. Dekker and Jan G. Verwer. *Stability of Runge-Kutta Methods for Stiff Nonlinear Differential Equations*. Elsevier Science Ltd., 1984.

Chapter 8

Partial Differential Equations

8.1 Introduction

Partial differential equations are ubiquitous in describing a large number of science and engineering problems, where the dependent variable is a function of more than one independent variable. For instance, in unsteady (or transient) heat diffusion in a two-dimensional solid domain, the temperature is a function of time and location. The governing energy conservation equation can be written in Cartesian coordinate system as

$$\rho c_p \frac{\partial T}{\partial t} = \frac{\partial}{\partial x}\left(k\frac{\partial T}{\partial x}\right) + \frac{\partial}{\partial y}\left(k\frac{\partial T}{\partial y}\right). \tag{8.1}$$

In general, in a two-independent variable problem, the second-order partial differential equation can be written as

$$A(x,y)\frac{\partial^2 f}{\partial x^2} + B(x,y)\frac{\partial^2 f}{\partial x \partial y} + C(x,y)\frac{\partial^2 f}{\partial y^2} = F\left(x,y,f,\frac{\partial f}{\partial x},\frac{\partial f}{\partial y}\right). \tag{8.2}$$

If

- $B - 4AC = 0$, the equation is parabolic.
- $B - 4AC < 0$, the equation is elliptic.
- $B - 4AC > 0$, the equation is hyperbolic.

The above classification has importance in selecting the numerical method to solve the equation with. Also, it relates to the physics of the problem.

Moreover, we can classify the equation based on the physics of the problem, e.g., equilibrium, diffusion, and advection. For example, in the elliptic equation (also called equilibrium or boundary-dictated problem), the information instantly "propagates" into the domain. Hence, any change in the boundary affects the whole domain values instantly. There is no initial condition to be applied to elliptic problems.

Unsteady advection-diffusion problems and unsteady diffusion problems are parabolic in nature. The information advects through and diffuses into the domain as time progresses. Both initial and boundary conditions have to be provided.

Wave equation is hyperbolic. The solution is not only dependent on time but also on velocity and, maybe, on acceleration.

In the following chapters, each of those types of partial differential equations will be discussed and the way to solve them numerically is presented.

8.2   Appropriateness of a Numerical Method

Let P be an operator defined in a given domain D. From a practical standpoint, we do consider here operators that represent a partial differential equation or sets of partial differential equations and their associated boundary and initial conditions. Let P_h a projection of P onto a finite countable subspace D_h. D_h is a finite partition of D, i.e., a set of non-overlapping parts (cells or elements or even sub-volumes) of D whose union is D. In other words, D_h is the mesh associated with the domain D. h is an indicator of the size (or volume) of these parts. Such a mesh is given in Fig. 8.1. P_h is the discrete problem associated with the continuous problem P. P_h can be based on any numerical method, e.g., FD or FV method.

An important question arises: Is the solution to the problem P_h a "good approximation" of the exact solution to the problem P? In other words, does the solution to the problem P_h converges to the (restriction of the) solution to the problem P (to D_h) when $h \to 0$?

8.2.1   Lax–Richtmyer theorem

By analogy with ordinary differential equations (see Section 6.8.4 of Chapter 6), the answer to the previous question — for linear initial

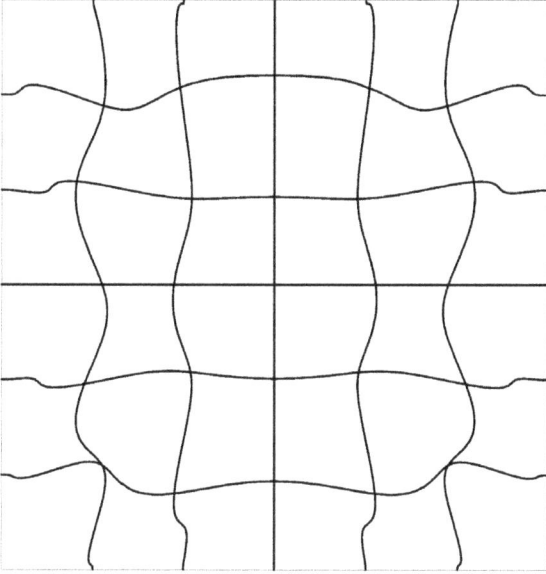

Figure 8.1: A two-dimensional non-uniform mesh. The size h can be the mean area of the mesh elements, that is $\frac{S_D}{N}$, where S_D is the area of the domain D and N the number of elements.

value partial differential equations — is "Yes", if the following conditions (Lax–Richtmyer theorem, once more) are satisfied:

1. The continuous problem P must be a well-posed problem. A problem is said to be well-posed if it admits a unique solution that shows *small* changes if the boundary and initial conditions are changed *slightly*. Lax–Milgram theorem, not presented here for the sake of conciseness, is a powerful tool to arrive at said proof as far as linear differential equations are concerned.

2. The discrete problem P_h is consistent, that is, the restriction (or projection) on D_h of the solution to the problem P (given on D) must also be a solution to the problem P_h when $h \to 0$.

3. The discrete problem P_h is stable, i.e., its solution depends continuously on the boundary and initial conditions and a small change in these conditions leads to a small change in the solution.

It is worth recalling that the usefulness of this theorem lies in the fact that, while convergence of the numerical solution is usually difficult to establish, consistency and stability of the scheme associated with it are typically much easier to show. Hence, convergence is ordinarily ascertained through the Lax–Richtmyer theorem by assessing the consistency and stability.

8.2.2   Example of existence and uniqueness of the solution

Consider the one-dimensional Laplace equation:

$$\frac{d^2\varphi}{dx^2} = 0, \quad 0 \leqslant x \leqslant L, \tag{8.3}$$

where φ is the unknown, x the independent variable, and L the width of the domain. This problem admits the following general solution:

$$\varphi(x) = Bx + C, \tag{8.4}$$

where B and C are constants to be determined using the boundary conditions.

- With non-homogeneous Neumann boundary conditions at both ends, the problem is ill posed. In fact, if we set $\frac{d\varphi}{dx}(x = 0) = q_1$ and $\frac{d\varphi}{dx}(x = L) = q_2$, then:
  - When $q_1 = q_2$, the problem admits an infinite number of solutions ($\varphi(x) = q_1 x + C$, where C becomes here any arbitrary constant).
  - When $q_1 \neq q_2$, the problem admits no solution.

- With at least one Dirichlet boundary condition, the same problem becomes well posed. In fact:
  - If we set $\varphi(x = 0) = \alpha$ and $\varphi(x = L) = \beta$, then the problem admits the following unique solution: $\varphi(x) = \frac{\beta - \alpha}{L}x + \alpha$.
  - If we set $\varphi(x = 0) = \alpha$ and $\frac{d\varphi}{dx}(x = L) = q_1$, then the problem admits the following unique solution: $\varphi(x) = q_1 x + \alpha$.

The consistency and stability of ordinary differential equations were discussed in the previous chapters. For completeness, the following section also touches upon the topic for partial differential equations.

8.3   Consistency Analysis

The consistency of a numerical scheme is the measure of the difference between the solution to the continuous partial differential equation and the solution to its discretization. The key to such an analysis is the Taylor series, as mentioned in the previous chapter. Let us illustrate these concepts through an example.

When applying the scheme called MacCormack scheme (will be discussed in detail later on) to a linear advection equation,

$$\frac{\partial u}{\partial t} + c\frac{\partial u}{\partial x} = 0, \tag{8.5}$$

where c is a positive constant, we get the following discretized equation:

$$u_i^{n+1} = \frac{1}{2\Delta x^2}\left(c\Delta t(c\Delta t + \Delta x)u_{i-1}^n\right.$$
$$\left. - (c\Delta t - \Delta x)\left[2(c\Delta t + \Delta x)u_i^n - c\Delta t u_{i+1}^n\right]\right). \tag{8.6}$$

Here, u_i^n is the approximation of u at t^n and x_i, Δt is the time step, and Δx is the constant space step.

Taylor series expansions of u_i^{n+1}, u_{i+1}^n, and u_{i-1}^n about (x_i, t^n), respectively, read

$$u_i^{n+1} = u_i^n + \left.\frac{\partial u}{\partial t}\right|_i^n \Delta t + \left.\frac{\partial^2 u}{\partial t^2}\right|_i^n \frac{\Delta t^2}{2} + O(\Delta t^3,)$$

$$u_{i+1}^n = u_i^n + \left.\frac{\partial u}{\partial x}\right|_i^n \Delta x + \left.\frac{\partial^2 u}{\partial x^2}\right|_i^n \frac{\Delta x^2}{2} + O(\Delta x^3), \tag{8.7}$$

$$u_{i-1}^n = u_i^n - \left.\frac{\partial u}{\partial x}\right|_i^n \Delta x + \left.\frac{\partial^2 u}{\partial x^2}\right|_i^n \frac{\Delta x^2}{2} + O(\Delta x^3).$$

Introducing these expansions into Eq. (8.5), then rearranging and dividing by Δt yields

$$\left.\frac{\partial u}{\partial t}\right|_i^n + c\left.\frac{\partial u}{\partial x}\right|_i^n + O(\Delta t) + O(\Delta x) = 0. \tag{8.8}$$

So, the solution (8.6) approaches the solution to Eq. (8.5) when both Δt and Δx approach zero. In this specific case, the scheme is said to be consistent up to the first order in time and space. Note that the procedure is the same as the one we used in analyzing the ordinary differential equations, see Section 6.8.2.

8.4 Stability Analysis

The stability of the numerical schemes used in solving ordinary differential equations were discussed in Section 6.8.3 of Chapter 6. The concept of stability is the same for schemes applied to solve partial differential equations. However, in this section, we discuss the cause of stability in general and present a systematic analysis to approach it.

In computers, data are stored in groups of eight digits (zeros and ones) called bytes. Consequently, computers never deal with real numbers themselves. For instance, the exact value of irrational number π is definitely unreachable to the computers, i.e., the value must be truncated at some point. Instead, any real number is represented using the floating point representation. This representation reads $Y = a \times b^n$, where a is called the mantissa, b the base (usually 2, 10, or 16), and n the exponent. For a representation of a real number using four bytes, for instance, one byte is used for the exponent absolute value and sign, one digit for its sign and 23 digits for its absolute value. So, in four-byte representation, π value reads $+0.31415927 \times 10^{+1}$. The smallest non-zero positive real number is therefore $\simeq 10^{-39}$. The consequence of such a finite representation is the round-off error that arises each time an arithmetic operation (addition, multiplication, etc.) is performed.

In time marching–like methods (thus involving parabolic or hyperbolic PDEs), the problem can always be cast in the following form:

$$\mathbf{A}\mathbf{u}^{n+1} = \mathbf{B}\mathbf{u}^n + \mathbf{C}, \qquad (8.9)$$

where $\mathbf{A}$ and $\mathbf{B}$ are constant matrices for linear problems and $\mathbf{C}$ a vector that does not depend on the solution $\mathbf{u}^n$. Vectors $\mathbf{u}^n$ and $\mathbf{u}^{n+1}$ are the vectors of the solution at time steps n and $n+1$, respectively. Their elements are the values of the solution over the meshed domain. At this step, vector $\mathbf{u}^n$ is assumed already determined, and the algebraic equation (8.9) allows us to retrieve vector $\mathbf{u}^{n+1}$. Note that $\mathbf{u}^n$ itself was determined using (8.9) by replacing n by $n-1$.

The starting point of stability analysis is as follows. Let $\mathbf{u}_b^n$ be a (base) solution to Eq. (8.9) and $\delta\mathbf{u}_b^n$ an infinitesimal perturbation of solution $\mathbf{u}_b^n$. The quantity $\mathbf{u}^n + \delta \mathbf{u}_b^n$ is a solution to Eq. (8.9), that is,

$$\mathbf{A}\left(\mathbf{u}_b^{n+1} + \delta \mathbf{u}^{n+1}\right) = \mathbf{B}\left(\mathbf{u}_b^n + \delta \mathbf{u}^n\right) + \mathbf{C}. \tag{8.10}$$

Hence, $\delta \mathbf{u}_b^n$ satisfies the equation

$$\delta \mathbf{u}^{n+1} = \left(\mathbf{A}^{-1}\mathbf{B}\right)\delta \mathbf{u}^n. \tag{8.11}$$

Note that in the above equation, the vector $\mathbf{C}$ cancels. (This explains why function $F(x)$ in the example presented in Section 6.8.3 of Chapter 6 can be set to zero without affecting the stability analysis outcome.) The gain G (or amplification factor) is defined by the ratio $G = \frac{\|\delta \mathbf{u}^{n+1}\|}{\|\delta \mathbf{u}^n\|}$. So, in order for the disturbance to not grow when n increases, the numerical method (that allowed us to write the matrices $\mathbf{A}$ and $\mathbf{B}$) is stable if and only if $|G| \leqslant 1$ or, equivalently, if the spectral radius of matrix $\mathbf{A}^{-1}\mathbf{B}$ is less than or equal to unity. (See Section 2.2 for the definition of the spectral radius.) This forms the stability criterion associated with the numerical method applied to a given PDE.

In many cases, determining the spectral radius of matrix $\mathbf{A}^{-1}\mathbf{B}$ is a cumbersome task. Fortunately, von Neumann suggested an analysis based on normal mode decomposition that is more tractable. For illustrative purpose, let us consider once again the MacCormack scheme applied to the advection equation (8.6). We use uniform mesh, $x_i = i\Delta x$, and u_i^n is the solution at (x_i, t^n), i.e., it is the ith element of vector $\mathbf{u}^n$. Similarly, δu_i^n is the disturbance in u_i^n.

von Neumann analysis expands the disturbance δu_i^n using Fourier series (or normal modes), that is,

$$\delta u_i^n = \sum_{k=-\infty}^{\infty} M_k \left[G(k)\right]^n \exp(Iki\Delta x). \tag{8.12}$$

Here, $I^2 = -1$, k is a wave number, $G(k)$ is the gain (function of the wave number k), and M_k is the Fourier series coefficient of δu_i^n. (Read the exponent n in $[G(k)]^n$ as a power and not as a superscript.) If the system is linear, then the analysis can be simplified by writing the disturbance as

$$\delta u_i^n = M\left[G(k)\right]^n \exp(Iki\Delta x). \tag{8.13}$$

As before, in order for the scheme to be stable, the magnitude of the ratio $\frac{\delta u_i^{n+1}}{\delta u_j^n}$ $(=|G|)$ must be such that

$$\forall k: \quad |G| \leqslant 1. \tag{8.14}$$

This is the new shape of the stability criterion. von Neumann analysis assumes the boundary conditions compatible with Eq. (8.9), that is, periodic boundary conditions. Otherwise, criterion (8.14) is to be considered as a necessary and not, in general, sufficient stability condition.

8.4.1   A first example

As an example, von Neumann stability analysis of MacCormack scheme (8.6) reads

$$MG^{n+1}e^{Iki\Delta x} = \frac{c\Delta t}{2\Delta x^2}(c\Delta t + \Delta x)MG^n e^{Ik(i-1)\Delta x}$$

$$- \frac{2(c\Delta t - \Delta x)(c\Delta t + \Delta x)}{2\Delta x^2}MG^n e^{Iki\Delta x}$$

$$+ \frac{c\Delta t - \Delta x}{2\Delta x^2}MG^n e^{Ik(i+1)\Delta x}. \tag{8.15}$$

By dividing both sides by $MG^n e^{Iki\Delta x}$, setting $k\Delta x = \theta$, recalling Euler's formula $(\exp(I\theta) = \cos\theta + I\sin\theta)$ and rearranging, we get

$$G = 1 + \left(\frac{c\Delta t}{\Delta x}\right)^2(-1 + \cos\theta) + I\frac{c\Delta t}{\Delta x}\sin\theta. \tag{8.16}$$

The gain G here is complex, so the stability condition reads $\forall\theta:$ $|G|^2 \leqslant 1$, *viz.*:

$$\left(1 + \left(\frac{c\Delta t}{\Delta x}\right)^2(-1 + \cos\theta)\right)^2 + \left(\frac{c\Delta t}{\Delta x}\right)^2(1 - \cos^2\theta) \leqslant 1, \tag{8.17}$$

or equivalently, by noticing that $(1 - \cos\theta) \geqslant 0$,

$$\frac{c\Delta t}{\Delta x} \leqslant 1. \tag{8.18}$$

The quantity $\frac{c\Delta t}{\Delta x}$ is usually called the Courant–Frederich–Lewy (CFL) number (or simply Courant number) and the stability criterion of MacCormack scheme is merely CFL $\leqslant 1$.

8.4.2 A second example

Another example to explain von Neumann stability analysis is as follows. As mentioned before, the method is based on Fourier series expansion, any periodic function can be expanded using Fourier series:

$$f(x,t) = \sum_{m=-\infty}^{\infty} A_m e^{Ik_m(x)}, \qquad (8.19)$$

where $k_m = 2m\pi/(2L)$ is the wave number and $I^2 = -1$. A_m is the amplitude at time level n. For instance,

$$f_i^n = A_m e^{Ik_m(i\Delta x)}, \qquad (8.20)$$

$$f_{i+1}^n = A_m e^{Ik_m(i+1)\Delta x} = f_i^n e^{Ik_m\Delta x}, \qquad (8.21)$$

and so on.

For example, let us consider one-dimensional diffusion equation:

$$\frac{\partial f}{\partial t} = \lambda \frac{\partial^2 f}{\partial x^2}. \qquad (8.22)$$

The above equation can be approximated using forward difference in time and central difference in space, i.e., by considering the explicit scheme,

$$\frac{f_i^{n+1} - f_i^n}{\Delta t} = \lambda \frac{f_{i+1}^n - 2f_i^n + f_{i-1}^n}{\Delta x^2} \qquad (8.23)$$

or

$$f_i^{n+1} = f_i^n(1 - 2w) + w(f_{i+1}^n + f_{i-1}^n), \qquad (8.24)$$

where $w = \lambda \delta t/\delta x^2$ is called the cell Fourier number. Using Fourier expansion,

$$f_i^{n+1} = f_i^n(1 - 2w) + w(f_i^n e^{Ik_m\Delta x} + f_i^n e^{-Ik_m\Delta x}). \qquad (8.25)$$

Since the amplification factor is the ratio of f_i^{n+1} to f_i^n, G, hence

$$G = \frac{f_i^{n+1}}{f_i^n} = (1 - 2w) + w(e^{I\theta} + e^{-I\theta}), \qquad (8.26)$$

where $\theta = k_m \Delta x$. Since $\cos \theta = \frac{e^{I\theta} + e^{-I\theta}}{2}$, the above equation becomes

$$G = 1 + 2w(\cos \theta - 1). \qquad (8.27)$$

For stable solution, $|G| \leqslant 1$ for all θ. Hence,

$$-1 \leqslant 1 - 2w(1 - \cos \theta) \leqslant 1. \qquad (8.28)$$

The maximum value of $1 - \cos \theta = 2$; therefore, $1 - 4w \geqslant -1$, $w \leqslant 1/2$.

In the following chapters, solutions of different types of differential equations will be elaborated.

Chapter 9

Diffusion Equation
(Parabolic Equation)

9.1   Introduction

Many processes in engineering and nature are governed by the diffusion equation, e.g., heat diffusion through a wall exposed to solar irradiance, water seepage through the soil, and pharmaceutical drugs diffusion into the living cells. The unsteady one-dimensional Cartesian coordinate diffusion equation can be written as

$$\frac{\partial f}{\partial t} = \frac{\partial}{\partial x}\left(\lambda \frac{\partial f}{\partial x}\right), \tag{9.1}$$

where λ is the diffusion coefficient (or diffusivity). For a constant λ,

$$\frac{\partial f}{\partial t} = \lambda \frac{\partial^2 f}{\partial x^2}. \tag{9.2}$$

For instance, for heat diffusion, $\lambda = \frac{k}{\rho c}$, where k, ρ, and c are thermal conductivity, density, and specific heat, respectively. Since the above equation has first-order derivative in time and second-order derivative in space, an initial value and two boundary values are required to close the solution.

There are many techniques to solve the diffusion equation.

179

9.2   Finite Difference

9.2.1   Explicit methods

A simple method is to approximate the time derivative using first-order forward difference (FD) and the space derivative using second-order central difference:

$$\frac{f_i^{n+1} - f_i^n}{\Delta t} = \lambda \frac{f_{i+1}^n - 2f_i^n + f_{i-1}^n}{\Delta x^2}, \tag{9.3}$$

where the superscript n stands for time step. The above equation is an explicit scheme. The only unknown in this equation is f_i^{n+1}:

$$f_i^{n+1} = (1 - 2w)f_i^n + wf_{i+1}^n + wf_{i-1}^n, \tag{9.4}$$

where $w = \lambda \Delta t / \Delta x^2$.

Note: All the explicit schemes are conditionally stable at best. It is clear that for stability condition, w should be equal or less than $1/2$. The stability of this scheme was discussed in Chapter 8. However, as a rule of thumb, the coefficients of the term f_i^n should be positive, otherwise the scheme is unstable. It is clear that the coefficient of f_i^n will be negative if the term $1 - 2w$ becomes negative. Hence, w should be equal or less than $1/2$ to ensure the stability. Recall that von Neumann analysis also leads to the same conclusion.

The scheme is first-order accurate in time and second-order accurate in space.

The reader may think that, being first-order accurate in time, the explicit method we introduced can be improved by using central difference to approximate the time derivative, i.e., $\frac{\partial f}{\partial t} = \frac{f_i^{n+1} - f_i^{n-1}}{2\Delta t}$; such an attempt was already made (called Richardson or leapfrog method) and the scheme so obtained is unconditionally unstable. Let us show that. The FD scheme is

$$\frac{f_i^{n+1} - f_i^{n-1}}{2\Delta t} = \lambda \frac{f_{i+1}^n - 2f_i + f_{i-1}^n}{\Delta x^2}. \tag{9.5}$$

Rearranging gives

$$f_i^{n+1} = f_i^{n-1} + w(f_{i+1}^n - 2f_i^n + f_{i-1}^n), \tag{9.6}$$

where $w = \frac{2\lambda\Delta t}{\Delta x^2}$. It is clear that the coefficient of f_i^n is negative regardless of the value of w. von Neumann method can be used to prove the unconditional instability of the scheme. This is done in Section 9.3.

However, a slight modification done to the above scheme can improve the stability condition by averaging in time the term f_i, the scheme is called DuFort–Frankel method:

$$\frac{f_i^{n+1} - f_i^{n-1}}{2\Delta t} = \lambda \frac{f_{i+1}^n - 2\left(\frac{f_i^{n+1}+f_i^{n-1}}{2}\right) + f_{i-1}^n}{\Delta x^2}. \tag{9.7}$$

Hence,

$$f_i^{n+1} = \frac{f_i^{n-1}(1-2w) + 2w(f_{i+1}^n + f_{i-1}^n)}{1+2w}. \tag{9.8}$$

The scheme is second-order accurate in time and in space. It is interesting that the scheme is unconditionally stable, but it is inconsistent. Another drawback of a scheme that is second-order accurate in time is that the value of f_i^{n-1} is required to start the solution. A second initial condition has to be provided, while physics offers only one.

Let us analyze the consistency of the scheme by using Taylor series expansion:

$$(1+2w)\left[f_i^n + \frac{\partial f}{\partial t}\Delta t + \frac{\partial^2 f}{\partial t^2}\frac{\Delta t^2}{2} + \cdots\right]$$

$$= (1-2w)\left[f_i^n - \frac{\partial f}{\partial t}\Delta t + \frac{\partial^2 f}{\partial t^2}\frac{\Delta t^2}{2} + \cdots\right]$$

$$+ 2w\left[\left(f_i^n + \frac{\partial f}{\partial x}\Delta x + \frac{\partial^2 f}{\partial x^2}\frac{\Delta x^2}{2} + \cdots\right)\right.$$

$$+ \left.\left(f_i^n - \frac{\partial f}{\partial x}\Delta x + \frac{\partial^2 f}{\partial x^2}\frac{\Delta x^2}{2} - \cdots\right)\right].$$

Rearranging the above equation gives

$$2\frac{\partial f}{\partial t} + 2w\frac{\partial^2 f}{\partial t^2}\Delta t = \frac{2w}{\Delta t}\frac{\partial^2 f}{\partial x^2}\Delta x^2 + \cdots \tag{9.9}$$

or

$$\frac{\partial f}{\partial t} = \lambda\frac{\partial^2 f}{\partial x^2} - \frac{\lambda\Delta t^2}{\Delta x^2}\frac{\partial^2 f}{\partial t^2}. \tag{9.10}$$

It is clear that as Δx approaches zero, the original equation is not covered. We can manipulate the last right-hand term by differentiation of the original equation with respect of time, hence

$$\frac{\partial}{\partial t}\left(\frac{\partial f}{\partial t}\right) = \lambda\frac{\partial}{\partial t}\left(\frac{\partial^2 f}{\partial x^2}\right) = \lambda\frac{\partial^2}{\partial x^2}\left(\frac{\partial f}{\partial t}\right) = \lambda^2\frac{\partial^4 f}{\partial x^4}. \tag{9.11}$$

Hence, Eq. (9.10) can be written as

$$\frac{\partial f}{\partial t} = \lambda\frac{\partial^2 f}{\partial x^2} - \frac{\lambda^3\Delta t^2}{\Delta x^2}\frac{\partial^4 f}{\partial x^4}. \tag{9.12}$$

The leading error is numerical diffusion.

Method of lines: The partial differential equation can be transformed into a system of ordinary, initial value problems by approximating the differential equation in space only, i.e.,

$$\frac{df_i}{dt} = \lambda\frac{f_{i+1} - 2f_i + f_{i-1}}{\Delta x^2}. \tag{9.13}$$

Such an approximation leads to a system of ordinary differential equations (ODEs), which can be solved using any method presented in Chapter 7, like the fourth-order Rung–Kutta method. The scheme is called method of lines. Hence, the scheme will be fourth order in time and second order in space. However, at each time step, there is a need to solve a system of ODEs using the fourth-order RK method. For example, if the domain of interaction is divided into N segments in space ($x-$ direction), with Dirichlet boundary conditions, $N - 1$ ODEs have to be solved.

9.2.2   Implicit methods

If the time derivative is evaluated at updated step, the scheme is called implicit scheme:

$$\frac{f_i^{n+1} - f_i^n}{\Delta t} = \lambda\frac{f_{i+1}^{n+1} - 2f_i^{n+1} + f_{i-1}^{n+1}}{\Delta x^2}. \tag{9.14}$$

The above equation can be written as

$$wf_{i-1}^{n+1} - (1 + 2w)f_i^{n+1} + wf_{i+1}^{n+1} = -f_i^n. \tag{9.15}$$

Hence, a system of algebraic equations forms a tri-diagonal matrix, which can be solved using Thomas algorithm, see Section 2.4.2 of

Chapter 2. The scheme is unconditionally stable. The accuracy of the solution is determined by the time step and space step size.

9.2.3 Crank–Nicolson method

Crank–Nicolson method is based on taking the time derivative at the center point between two time steps:

$$\frac{f_i^{n+1} - f_i^n}{\Delta t} = \frac{\lambda}{2} \left[\frac{f_{i+1}^{n+1} - 2f_i^{n+1} + f_{i-1}^{n+1}}{\Delta x^2} + \frac{f_{i+1}^n - 2f_i^n + f_{i-1}^n}{\Delta x^2} \right].$$

$$(9.16)$$

The scheme is second-order accurate in time and space. The approximation leads to a tri-diagonal matrix. The scheme is conditionally stable. However, the stability condition is much better than the explicit scheme.

For problems with source term, i.e.,

$$\frac{\partial f}{\partial t} = \lambda \frac{\partial^2 f}{\partial x^2} + S,$$

$$(9.17)$$

the source term S need to evaluated as $\frac{1}{2}(S_i^n + S_i^{n+1})$.

9.3 Unstable and Inconsistent Schemes

Let us analyze more deeply the second-order time-accurate scheme presented in Section 9.2.1 that was suggested by Richardson in 1910. The mentioned scheme reads

$$\frac{f_i^{n+1} - f_i^{n-1}}{2\Delta t} = \lambda \frac{f_{i+1}^n - 2f_i^n + f_{i-1}^n}{\Delta x^2}.$$

$$(9.18)$$

The above scheme is second-order accurate both in time and space. Also, the scheme is consistent. **However, it is unconditionally unstable**. To prove the above statements, let us rearrange the above equation as

$$f_i^{n+1} - f_i^{n-1} = \beta(f_{i+1}^n - 2f_i^n + f_{i-1}^n),$$

$$(9.19)$$

where $\beta = \frac{2\lambda \Delta t}{\Delta x^2}$.

Using Taylor series expansion, the left-hand side reads (for better clarity, the index i is omitted in the derivatives f_t, f_{tt}, etc.)

$$f_i + f_t \Delta t + f_{tt} \Delta t/2! + f_{ttt} \Delta t/3! + O(\Delta t^4)$$
$$- (f_i - f_t \Delta t + f_{tt} \Delta t/2! - f_{ttt} \Delta t/3! + O(\Delta t^4)), \quad (9.20)$$

which ends up as

$$2 f_t \Delta t + f_{ttt} \Delta t^3/3 + O(\Delta t^5). \quad (9.21)$$

Applying Taylor series for the right-hand side gives

$$f_i + f_x \Delta x + f_{xx} \Delta x^2/2! + O(\Delta x^3)$$
$$- 2 f_i + (f_i - f_x \Delta x + f_{xx} \Delta x^2/2! + O(\Delta x^3)), \quad (9.22)$$

which ends up as

$$f_{xx} \Delta x^2 + O(\Delta x^4). \quad (9.23)$$

Hence, the scheme reads

$$2 f_t \Delta t + f_{ttt} \Delta t^3/3 + O(\Delta t^5) = \beta \left(f_{xx} \Delta x^2 + O(\Delta x^4) \right). \quad (9.24)$$

Dividing by $2\Delta t$ and rearranging yields

$$f_t = \lambda f_{xx} - f_{ttt} \Delta t^2/3 + O(\Delta x^2). \quad (9.25)$$

Hence, the scheme is consistent and is second-order accurate in time and space indeed.

Let us use von Neumann stability analysis, where $f_{i+1} = f_i e^{Ik_m \Delta x}$ and $f_{i-1} = f_i e^{-Ik_m \Delta x}$.

Equation (9.19) is substituted by

$$f_i^{n+1} = f_i^{n-1} + \beta(f_i^n e^{Ik_m \Delta x} - 2 f_i^n + f_i^n e^{-Ik_m \Delta x}). \quad (9.26)$$

Dividing the equation by f_i and letting $Ik_m \Delta x = \theta$ yields

$$\frac{f_i^{n+1}}{f_i^n} = \frac{f_i^{n-1}}{f_i^n} + \beta(e^\theta - 2 + e^{-I\theta}). \quad (9.27)$$

Note that $G = \frac{f_i^{n+1}}{f_i^n} = \frac{f_i^n}{f_i^{n-1}}$ (amplification factor). Recall that $\cos\theta = \frac{e^\theta + e^{-\theta}}{2}$. So,

$$G = \frac{1}{G} + 2\beta(\cos\theta - 1), \qquad (9.28)$$

or

$$G^2 - 2\beta(\cos\theta - 1)G - 1 = 0, \qquad (9.29)$$

which is a quadratic equation and its solution is

$$G = \frac{-b \pm \sqrt{b^2 + 4}}{2}, \qquad (9.30)$$

where $b = 2\beta(\cos\theta - 1)$ and $\beta = \frac{2\lambda\Delta t}{\Delta x^2}$. Hence,

$$G = -\beta(\cos\theta - 1) \pm \sqrt{\beta^2(\cos\theta - 1)^2 + 1}. \qquad (9.31)$$

The stability condition reads $\forall\theta \in [0; 2\pi) : |G| \leqslant 1$. Note though, since $\lambda\Delta t \neq 0$, that $|G| = 1$ only for $\theta = 0$ and $|G| > 1$ for all other values of θ. So, the scheme is unconditionally unstable.

To remedy the stability problem, DuFort–Frankel amended scheme can be applied. It consists of replacing f_i^n by $(f_i^{n+1} + f_i^{n-1})/2$. Hence, the final result becomes

$$(1 + 2\alpha)f_i^{n+1} = (1 - 2\alpha)f_i^{n-1} + 2\alpha(f_{i+1}^n + f_{i-1}^n), \qquad (9.32)$$

where $\alpha = \lambda\Delta t/\Delta x^2$. The stability analysis of this equation yields the following algebraic equation of the unknown G:

$$(1 + 2\alpha)G = \frac{1 - 2\alpha}{G} + 4\alpha\cos\theta. \qquad (9.33)$$

The solutions of this equation read

$$G_1 = \frac{2\alpha\cos\theta - sqrt{1 - 4\alpha^2 + 4\alpha^2\cos^2\theta}}{1 + 2\alpha},$$

$$G_2 = \frac{2\alpha\cos\theta + sqrt{1 - 4\alpha^2 + 4\alpha^2\cos^2\theta}}{1 + 2\alpha}. \qquad (9.34)$$

It can be shown that the maximum of $|G_1|$ and $|G_2|$ is reached for $\theta = 0$, in which case $|G_1| = \frac{2\alpha - 1}{2\alpha + 1} < 1$ and $|G_1| = 1$ for all α. The scheme is unconditionally stable indeed.

However, it is inconsistent, in virtue of the consistency analysis that yields

$$f_t = \lambda f_{xx} - \lambda f_{tt} \frac{\Delta t^2}{\Delta x^2} - \frac{1}{6} f_{ttt} \Delta t^3 + \frac{1}{12} \lambda f_{xxxx} \Delta x^2 + \cdots . \quad (9.35)$$

As Δx and Δt approach zero, the second term on the right-hand side does not vanish unless Δt approaches zero *faster* than Δx.

9.4   Multi-dimension

Consider a two-dimensional, unsteady, constant-coefficient diffusion equation in Cartesian coordinates:

$$\frac{\partial f}{\partial t} = \lambda \left(\frac{\partial^2 f}{\partial x^2} + \frac{\partial^2 f}{\partial y^2} \right) . \quad (9.36)$$

This equation can be approximated in a similar manner as previously. The explicit approximation reads

$$\frac{f_{i,j}^{n+1} - f_{i,j}^n}{\Delta t} = \lambda \left[\frac{f_{i+1,j}^n - 2f_{i,j}^n + f_{i-1,j}^n}{\Delta x^2} + \frac{f_{i,j+1}^n - 2f_{i,j}^n + f_{i,j-1}^n}{\Delta y^2} \right],$$
$$(9.37)$$

which has orders of accuracy of Δt, Δx^2, and Δy^2. Let us perform the stability analysis. Since the problem is two-dimensional, $f_{i,j}^n$ has to be expanded as follows:

$$f_{i,j}^n == M\,[G(k)]^n \exp(Iki\Delta x) \exp(Ik'j\Delta y), \quad (9.38)$$

where two independent disturbance modes, k and k', are involved. So, substituting $f_{i,j}^{n+1}$, $f_{i+1,j}^n$, $f_{i-1,j}^n$, $f_{i,j+1}^n$, and $f_{i,j-1}^n$, accordingly, into Eq. (9.37), putting $k\Delta x = \theta$, $k'\Delta y = \theta'$, and dividing by $f_{i,j}^n$, we retrieve

$$G = 1 + 2\frac{\lambda \Delta t}{\Delta x^2}(\cos\theta - 1) + 2\frac{\lambda \Delta t}{\Delta y^2}(\cos\theta' - 1). \quad (9.39)$$

So, for the scheme to be stable, the condition $\forall \theta, \theta' : -1 \leqslant G \leqslant 1$ must hold, which ends up as (since $\lambda \Delta t > 0$)

$$\forall \theta, \theta' \in [0; 2\pi) : 2\frac{\lambda \Delta t}{\Delta x^2}(1 - \cos\theta) + 2\frac{\lambda \Delta t}{\Delta y^2}(1 - \cos\theta') \leqslant 2. \quad (9.40)$$

$|G|$ being maximum for $\theta = \theta' = \pi$, the scheme is conditionally stable, and the stability condition reads

$$\lambda \Delta t \left(\frac{1}{\Delta x^2} + \frac{1}{\Delta y^2} \right) \leqslant \frac{1}{2}. \tag{9.41}$$

Now, let us examine the implicit scheme, which reads

$$\frac{f_{i,j}^{n+1} - f_{i,j}^{n}}{\Delta t} = \lambda \left[\frac{f_{i+1,j}^{n+1} - 2f_{i,j}^{n+1} + f_{i-1,j}^{n+1}}{\Delta x^2} + \frac{f_{i,j+1}^{n+1} - 2f_{i,j}^{n+1} + f_{i,j-1}^{n+1}}{\Delta y^2} \right], \tag{9.42}$$

which has the same order of accuracy as the explicit method. However, the scheme is unconditionally stable (left as an exercise). The above equation can be rewritten as

$$a_{i,j} f_{i+1,j}^{n+1} + b_{i,j} f_{i-1,j}^{n+1} + c_{i,j} f_{i,j}^{n+1} + d_{i,j} f_{i,j+1}^{n+1} + e_{i,j} f_{i,j-1}^{n+1} = -f_{i,j}^{n}, \tag{9.43}$$

where $a_{i,j} = (\lambda \Delta t)/\Delta x^2$, $b_{i,j} = (\lambda \Delta t)/\Delta x^2$, $c_{i,j} = -[1 + 2\lambda \Delta t (1/\Delta x^2 + 1/\Delta y^2)]$, $d_{i,j} = (\lambda \Delta t)/\Delta y^2$, and $e_{i,j} = (\lambda \Delta t)/\Delta y^2$. In order to solve efficiently (at a given time step) the above system of linear algebraic equations, a special approach needs to be selected. This is presented in the following section.

9.5   Alternating Direction Implicit Schemes

Two- and three-dimensional implicit schemes lead inevitably to sparse matrices. For large matrices, the convergence of the solution becomes slow, even when using iterative methods. Alternating direction implicit (ADI) schemes may help.

The key idea is to convert the resulting matrix into a tri-diagonal matrix by moving all elements to the right-hand side of the matrix equation, except the diagonal, upper diagonal, and lower diagonal elements. Recall the implicit formulation:

$$\frac{f_{i,j}^{n+1} - f_{i,j}^{n}}{\Delta t} = \lambda \left[\frac{f_{i+1,j}^{n+1} - 2f_{i,j}^{n+1} + f_{i-1,j}^{n+1}}{\Delta x^2} + \frac{f_{i,j+1}^{n+1} - 2f_{i,j}^{n+1} + f_{i,j-1}^{n+1}}{\Delta y^2} \right]. \tag{9.44}$$

The above equation is split into steps $(\Delta t/2)$ in order to complete updating the solution at time step Δt. For the first half time step, the spatial derivatives are implicit for all derivatives in x direction and explicit in y direction, then alternate the direction in the second step. The first step is to solve for $\Delta t/2$, the solution being denoted by a star $(*)$:

$$-w_x f^*_{i-1,j} + (1 - 2w_x)f^*_{i,j} - w_x f^*_{i+1,j}$$
$$= (1 + 2w_y)f^n_{i,j} + w_y(f^n_{i,j+1} + f^n_{i,j-1}), \qquad (9.45)$$

where w_x and w_y are $\lambda\Delta t/(2\Delta x^2)$ and $\lambda\Delta t/(2\Delta y^2)$, respectively. The second step, which completes updating solution at time step (Δt), reads

$$-w_y f^{n+1}_{i,j-1} + (1 - 2w_y)f^{n+1}_{i,j} - w_y f^{n+1}_{i,j+1}$$
$$= (1 + 2w_x)f^*_{i,j} + w_x(f^*_{i+1,j} + f^*_{i-1,j}). \qquad (9.46)$$

The scheme has orders of accuracy of $O(\Delta t)^2$, $O(\Delta x)^2$, and $O(\Delta y)^2$. A good strategy is to alternate the sweeping process between $x-$ and $y-$ directions. As shown in Fig. 9.1, the process of sweeping starts in $y-$ direction (implicit in $x-$ direction), then sweep in $x-$ direction

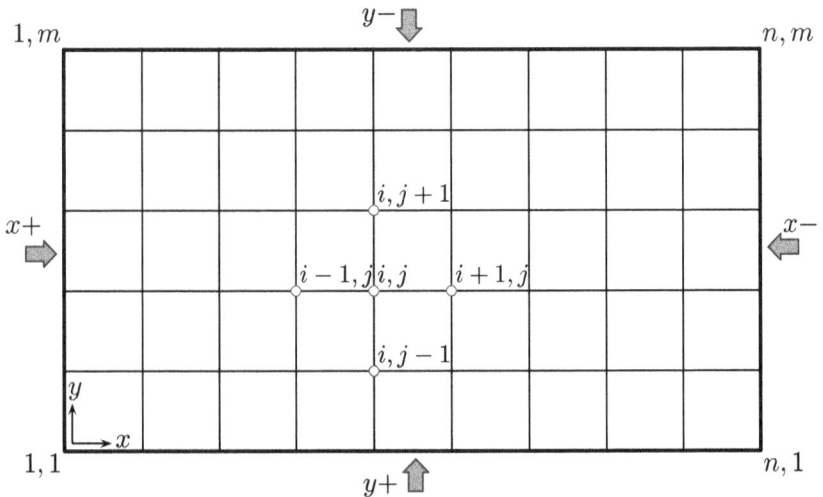

Figure 9.1:   Sweeping directions for ADI method.

(implicit in $y-$ direction). For the third sweep, it is advised to start in $y+$ direction (implicit in $x-$ direction), and in the forth sweeping, start sweeping in $x+$ (implicit in $y-$ direction). For the next time step, repeat the entire process.

9.6 Finite Volume Method

9.6.1 The key idea

The concept of the finite volume method (FVM) was already introduced when dealing with ODEs, see Section 7.6 in Chapter 7. This concept can be extended to partial differential equations in a straightforwad manner. For a two-dimensional, unsteady state diffusion equation with source term,

$$\frac{\partial f}{\partial t} = \frac{\partial}{\partial x}\left(\lambda \frac{\partial f}{\partial x}\right) + \frac{\partial}{\partial y}\left(\lambda \frac{\partial f}{\partial y}\right) + S, \tag{9.47}$$

the method is applied as follows. First, multiplying the above equation by a volume $(dt\,dx\,dy)$ and carrying out integration over time interval $[t;\,t+\Delta t]$ and control volume with faces w and e normal to the $x-$ direction and s and n normal to the $y-$ direction (see Fig. 9.2) yields

$$\int_w^e \int_s^n \int_t^{t+\Delta t} \frac{\partial f}{\partial t}\, dt\, dy\, dx = \int_t^{t+\Delta t} \int_s^n \int_w^e \frac{\partial}{\partial x}\left(\lambda \frac{\partial f}{\partial x}\right) dx\, dy\, dt$$

$$+ \int_t^{t+\Delta t} \int_w^e \int_s^n \frac{\partial}{\partial y}\left(\lambda \frac{\partial f}{\partial y}\right) dy\, dx\, dt$$

$$+ \int_t^{t+\Delta t} \int_s^n \int_w^e S\, dx\, dy\, dt.$$

Here, we assume that the mesh is such that the normal to each face is parallel to the line that joins the centers (also called centroids) of the two neighboring control volumes and that line passes though the face center. These are called non-skewed orthogonal meshes or grids. If they are not, special attention should be paid, but digging into such a consideration is outside the scope of this book.

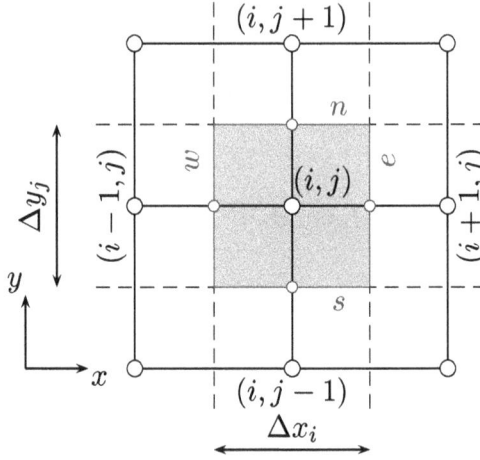

Figure 9.2:   Control volume. n, s, e, and x are its four faces.

Assuming that in the transient term function f is constant within the control volume, the above equation becomes

$$(f_{i,j}^{n+1} - f_{i,j}^n)\Delta x_i \Delta y_j = \left(\lambda_e \frac{\partial f}{\partial x}\bigg|_e - \lambda_w \frac{\partial f}{\partial x}\bigg|_w \right) \Delta y_j \Delta t$$

$$+ \left(\lambda_n \frac{\partial f}{\partial y}\bigg|_n - \lambda_s \frac{\partial f}{\partial y}\bigg|_s \right) \Delta x_i \Delta t$$

$$+ S_{i,j}\Delta t \Delta x_i \Delta y_j. \tag{9.48}$$

There are a few options for time updating of the right-hand side of the above equation: f can be taken explicit (evaluated at n), implicit (evaluated at $n+1$), or semi-implicit (evaluated at some in-between point where it can be replaced by $(f^{n+1} + f^n)/2$). The derivatives can be evaluated using finite difference or by passing a polynomial through a set of points (control volume centers). The value of the transport properties (λ) must be evaluated at the interface to ensure the flux continuity, hence the continuity of function f itself. This means that we must get the same value for f whether we approach the interface from the left side or from the right side. The second condition is the continuity of the derivative of f normal to the interface. For example, the derivative along the $x-$ direction is

$$\frac{\partial f}{\partial x}\bigg|_l = \frac{\partial f}{\partial x}\bigg|_r. \tag{9.49}$$

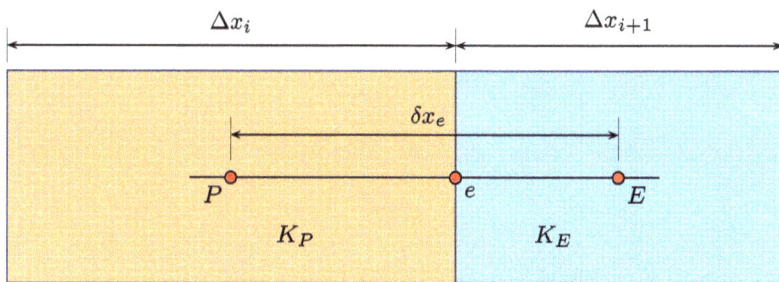

Figure 9.3:   Heat conduction through a two-media system.

The concept can easily be explained by considering heat conduction through two different media, as shown in Fig. 9.3, then by using the concept of thermal resistance.

The total thermal resistance between points P and E is the sum of the thermal resistances between P and e and between e and E. Therefore, the effective thermal conductivity can be calculated as

$$\frac{\delta x_e}{K_{\text{eff}}} = \frac{\Delta x_i}{2K_P} + \frac{\Delta x_{i+1}}{2K_E}, \tag{9.50}$$

where the following relationship holds: $\delta x_e = \dfrac{\Delta x_i + \Delta x_{i+1}}{2}$. Hence,

$$K_{\text{eff}} = \frac{2\delta x_e K_E K_P}{\Delta x_i K E + \Delta x_{i+1} K_P}. \tag{9.51}$$

Consequently, the flux reads

$$\left.\frac{\partial f}{\partial x}\right|_e = K_{\text{eff}} \frac{T_P - T_E}{\delta x_e}. \tag{9.52}$$

For explicit method, the right-hand side of Eq. (9.48) can be approximated as

$$\left(\lambda_e \frac{f_{i+1,j}^n - f_{i,j}^n}{\delta x_e} - \lambda_w \frac{f_{i,j}^n - f_{i-1,j}^n}{\delta x_w} \right) \Delta y_j \Delta t$$

$$+ \left(\lambda_n \frac{f_{i,j+1}^n - f_{i,j}^n}{\delta y_n} - \lambda_s \frac{f_{i,j}^n - f_{i,j-1}^n}{\delta y_s} \right) \Delta x_i \Delta t + S_{i,j} \Delta t \Delta x_i \Delta y_j. \tag{9.53}$$

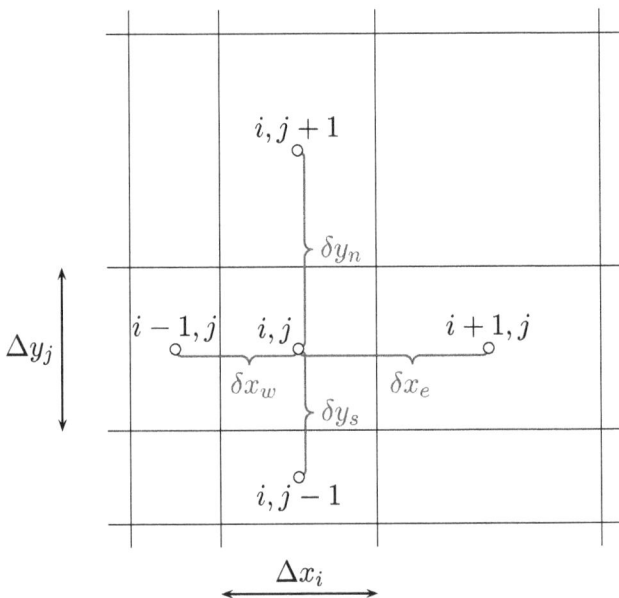

Figure 9.4: Control volume notations.

The terms are for nonuniform grids (volumes, see Fig. 9.4). Hence, the explicit FVM method yields

$$f_{i,j}^{n+1} = a_P f_{i,j}^n + a_E f_{i+1,j}^n + a_W f_{i-1,j}^n + a_N f_{i,j+1}^n + a_S f_{i,j-1}^n + S_{i,j}^n \Delta t, \tag{9.54}$$

where $a_E = \lambda_e \Delta t/(\delta x_e \Delta x_i)$, $a_W = \lambda_w \Delta t/(\delta x_w \Delta x_i)$, $a_N = \lambda_n \Delta t/(\delta y_n \Delta y_j)$, $a_S = \lambda_s \Delta t/(\delta y_s \Delta y_j)$, and $a_P = 1 - a_E - a_W - a_N - a_S$. It is clear that for stability condition, a_p must be positive, which implies that $(a_E + a_W + a_N + a_S)$ should be less than or equal to one.

For implicit scheme, all space derivatives are evaluated at time $t + \Delta t$, i.e., at $(n + 1)$. The ADI method can be used to solve the governing algebraic system of equations.

9.6.2   Boundary conditions for FVM

The simple and straightforward boundary condition is when the value of the function is given at the boundary. In the following section, other kinds of boundary conditions will be explained. For illustrative purpose, the west boundary condition will be taken as an example.

Adiabatic boundary condition: This boundary condition can be naturally applied by setting $\frac{\partial f}{\partial x}\big|_w$ to zero, that is, by canceling from the balance the flux at the west face.

Flux is given: This kind of boundary condition is naturally applied as well by setting $\frac{\partial f}{\partial x}\big|_w = q''$, where q'' is the given flux. Considering such a flux is equivalent to setting $\frac{\partial f}{\partial x}\big|_w$ to zero and adding the flux q'' as a source term to the control volume at the boundary. Since the source term is per unit volume in the governing equation, the added source term should read $q''/(\Delta x)$.

Third kind of boundary condition: As $\frac{\partial f}{\partial x}\big|_w = -\kappa(f - f_0)\big|_w$, where κ and f_0 are given constants. The best practice is to set $\frac{\partial f}{\partial x}\big|_w$ to zero (adiabatic condition) and add an extra source term to the control volume. To illustrate the process, let us assume that the boundary condition is the convective boundary condition, i.e., $h(T - T_0)\big|_w = -\lambda\frac{\partial T}{\partial x}\big|_w$. Using the concept of thermal resistance, $q'' = (T_0 - T_P)/R_t$, where T_0, T_p, and R_t are the known outside temperature, the control volume temperature, and total thermal resistance, respectively. $R_t = 1/h + \Delta x/(2\lambda)$, where $\Delta x/2$ is the distance between the boundary and T_P (the control volume center). The source term in the governing equation is per unit volume. Hence, the added source term should be $q''/\Delta x$, where Δx is the width of the control volume. Another way to consider such a boundary condition, which yields the same results, is presented in the following section.

9.7 Worked Example

Consider the following unsteady heat equation inside a spherical body of radius a with uniform properties (density, heat capacity, and heat conductivity):

$$\rho c_p \frac{\partial T}{\partial t} = \nabla \cdot (\lambda \nabla T), \tag{9.55}$$

and its initial and boundary conditions are

$$T = T_0 \quad \text{at } t = 0, \tag{9.56}$$

$$-\lambda \frac{\partial T}{\partial n} = h(\theta)(T - T_\infty) \quad \text{at } r = a, \tag{9.57}$$

where the heat exchange coefficient h is a function of the colatitude (or zenith angle) θ, as depicted in Fig. 9.5. T is the temperature, t the time, and r the distance from the center of the body. The geometry suggests the use of the spherical coordinates (r, θ, φ). Because the problem does not exhibit any explicit dependency on the meridianal angle φ, the temperature can be assumed to be a function of (r, θ, t) only. (In the case where thermal instabilities are analyzed, the solution can be sought as a function of θ too. But that's another story.)

In order to discretize the problem, consider Δr, $\Delta \theta$, and Δt, the radius, colatitude, and time step, respectively, and let $T_{i,j}^n$ be the temperature at some location (r_i, θ_j) at time t_n. We apply the FVM where projection into the plan (yOz) of the mesh is sketched

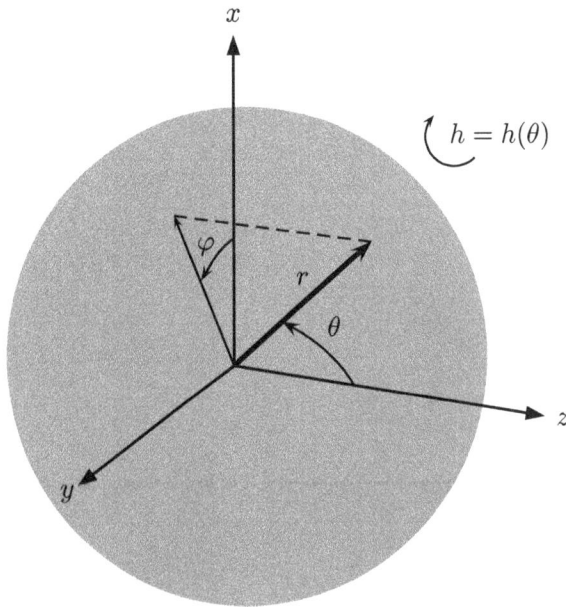

Figure 9.5:   Heat transfer problem inside a spherical body to be solved using the FVM.

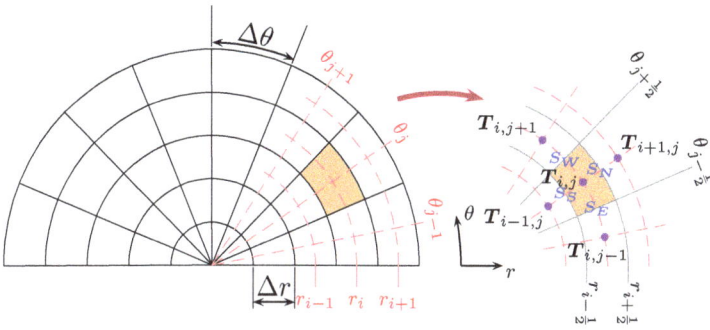

Figure 9.6: The FVM mesh (projected on the plan (xOz)) used to solve the heat transfer problem inside a spherical body. (For better clarity, the superscripts i, j are omitted in the faces' symbols.)

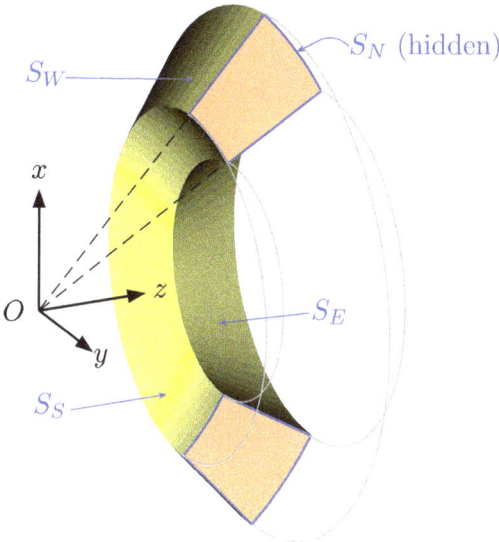

Figure 9.7: A toroid (truncated) representing a typical finite volume, a cell, $v_{i,j}$. (For better clarity, the superscripts i, j are omitted in the faces' symbols.)

in Fig. 9.6. So, each cell (or volume) $v_{i,j}$, illustrated in Fig. 9.7, is such that

$$r_{i-\frac{1}{2}} \leqslant r \leqslant r_{i+\frac{1}{2}} \quad i = 1, \ldots, m_r,$$

$$\theta_{j-\frac{1}{2}} \leqslant r \leqslant \theta_{j+\frac{1}{2}} \quad j = 1, \ldots, m_\theta, \tag{9.58}$$

$$0 \leqslant \varphi < 2\pi,$$

where m_r and m_θ are the number of cells in the r and θ directions, respectively. So, the space steps read

$$\Delta r = \frac{a}{m_r},$$

$$\Delta \theta = \frac{\pi}{m_\theta}.$$

According to Fig. 9.5, $\theta_{j-\frac{1}{2}} = \Delta\theta(j-1)$, $\theta_j = \frac{1}{2}\left(\theta_{j-\frac{1}{2}} + \theta_{j+\frac{1}{2}}\right) = \Delta\theta\left(j - \frac{1}{2}\right)$, $r_{i-\frac{1}{2}} = \Delta r(i-1)$, $r_{i+\frac{1}{2}} = \Delta r i$, and $r_i = \frac{1}{2}\left(r_{j-\frac{1}{2}} + r_{j+\frac{1}{2}}\right)$.
(Because of the complexity of the cell $v_{i,j}$ shape, the expression of r_i assumed here is only a first-order approximation; a higher-order approximation can be achieved but is not presented here for the sake of simplicity.)

By integrating Eq. (9.55) over the cell volume $v_{i,j}$, the latter does not depend on time, and applying the divergence theorem, we get

$$\rho c_p \frac{d}{dt} \int_{v_{i,j}} T\,dv = \sum_{k=N,S,E,W} \int_{S_k^{i,j}} \lambda \mathbf{n}_k \cdot \nabla T\,dS_k, \tag{9.59}$$

where $S_N^{i,j}$, $S_S^{i,j}$, $S_E^{i,j}$, and $S_W^{i,j}$ are the north, south, east, and west faces of the cell $v_{i,j}$, respectively, and $\mathbf{n}_k$ is the unit normal vector to face k, see Figs. 9.6 and 9.7. In spherical coordinates (r, θ, φ), $dv = r^2 \sin\theta\, dr\, d\theta\, d\varphi$, and $dS_N = r_{i+\frac{1}{2}}^2 \sin\theta\, d\theta\, d\varphi$, $dS_S = r_{i-\frac{1}{2}}^2 \sin\theta\, d\theta\, d\varphi$, $dS_E = r\sin\theta_{i-\frac{1}{2}} dr\, d\varphi$, and $dS_W = r\sin\theta_{i+\frac{1}{2}} dr\, d\varphi$. In addition, $\mathbf{n}_N = (1,0,0)^t$, $\mathbf{n}_S = (-1,0,0)^t$, $\mathbf{n}_E = (-1,0,0)^t$, and $\mathbf{n}_W = (1,0,0)^t$. Introducing these differentials into Eq. (9.59) yields

$$\rho c_p \frac{d}{dt} \int_{r=r_{i-\frac{1}{2}}}^{r_{i+\frac{1}{2}}} \int_{\theta=\theta_{j-\frac{1}{2}}}^{\theta_{j+\frac{1}{2}}} \int_{\varphi=0}^{2\pi} T\, r^2 \sin\theta\, dr\, d\theta\, d\varphi$$

$$= \sum_{k=N,S,E,W} \int_{S_k^{i,j}} \lambda \mathbf{n}_k \cdot \nabla T\,dS_k. \tag{9.60}$$

Assuming the temperature constant over the cell $v_{i,j}$, the left-hand side of Eq. (9.60) reads

$$\rho c_p \frac{dT_{i,j}}{dt} v_{i,j},$$

where

$$v_{i,j} = \int_{r=r_{i-\frac{1}{2}}}^{r_{i+\frac{1}{2}}} r^2 \, dr \times \int_{\theta=\theta_{j-\frac{1}{2}}}^{\theta_{j+\frac{1}{2}}} \sin\theta \, d\theta \times \int_{\varphi=0}^{2\pi} d\varphi$$

$$= \frac{2\pi}{3} \left(r_{i+\frac{1}{2}}^3 - r_{i-\frac{1}{2}}^3 \right) \left(\cos\theta_{j-\frac{1}{2}} - \cos\theta_{j+\frac{1}{2}} \right).$$

Assuming the temperature gradient constant over each face $S_k^{i,j}$, recall that $\nabla T = (\frac{\partial T}{\partial r}, \frac{1}{r}\frac{\partial T}{\partial\theta}, \frac{1}{r\sin\theta}\frac{\partial T}{\partial\varphi})^t$, the right-hand side of Eq. (9.60) reads

$$\lambda \left(\frac{\partial T}{\partial r} \right)_{r_{i+\frac{1}{2}},\theta_j} S_N^{i,j} - \lambda \left(\frac{\partial T}{\partial r} \right)_{r_{i-\frac{1}{2}},\theta_j} S_S^{i,j}$$

$$+ \frac{\lambda}{r_i} \left(\frac{\partial T}{\partial\theta} \right)_{r_i,\theta_{j+\frac{1}{2}}} S_W^{i,j} - \frac{\lambda}{r_i} \left(\frac{\partial T}{\partial\theta} \right)_{r_i,\theta_{j-\frac{1}{2}}} S_E^{i,j}.$$

where

$$S_N^{i,j} = \int_{\theta=\theta_{j-\frac{1}{2}}}^{\theta_{j+\frac{1}{2}}} \int_{\varphi=0}^{2\pi} dS_N = 2\pi r_{i+\frac{1}{2}}^2 \left(\cos\theta_{j-\frac{1}{2}} - \cos\theta_{j+\frac{1}{2}} \right).$$

$$S_S^{i,j} = \int_{\theta=\theta_{j-\frac{1}{2}}}^{\theta_{j+\frac{1}{2}}} \int_{\varphi=0}^{2\pi} dS_S = 2\pi r_{i-\frac{1}{2}}^2 \left(\cos\theta_{j-\frac{1}{2}} - \cos\theta_{j+\frac{1}{2}} \right).$$

$$S_W^{i,j} = \int_{r=r_{i-\frac{1}{2}}}^{r_{i+\frac{1}{2}}} \int_{\varphi=0}^{2\pi} dS_W = \pi \sin\theta_{j+\frac{1}{2}} \left(r_{i+\frac{1}{2}}^2 - r_{i-\frac{1}{2}}^2 \right),$$

$$S_E^{i,j} = \int_{r=r_{i-\frac{1}{2}}}^{r_{i+\frac{1}{2}}} \int_{\varphi=0}^{2\pi} dS_W = \pi \sin\theta_{j-\frac{1}{2}} \left(r_{i+\frac{1}{2}}^2 - r_{i-\frac{1}{2}}^2 \right)$$

and

$$\left(\frac{\partial T}{\partial r} \right)_{r_{i+\frac{1}{2}},\theta_j} = \frac{T_{i+1,j} - T_{i,j}}{\Delta r},$$

$$\left(\frac{\partial T}{\partial r} \right)_{r_{i-\frac{1}{2}},\theta_j} = \frac{T_{i,j} - T_{i-1,j}}{\Delta r},$$

$$\left(\frac{\partial T}{\partial \theta}\right)_{r_i, \theta_{j+\frac{1}{2}}} = \frac{T_{i,j+1} - T_{i,j}}{\Delta\theta},$$

$$\left(\frac{\partial T}{\partial \theta}\right)_{r_i, \theta_{j-\frac{1}{2}}} = \frac{T_{i,j} - T_{i,j-1}}{\Delta\theta}.$$

So, the final equation to be solved inside the domain, that is a cell having exactly four neighbors, reads

$$\rho c_p v_{i,j} \frac{dT_{i,j}}{dt} = \frac{\lambda S_N^{i,j}}{\Delta r}\left(T_{i+1,j} - T_{i,j}\right) - \frac{\lambda S_S^{i,j}}{\Delta r}\left(T_{i,j} - T_{i-1,j}\right)$$

$$+ \frac{\lambda S_W^{i,j}}{r_i \Delta\theta}\left(T_{i,j+1} - T_{i,j}\right) - \frac{\lambda S_E^{i,j}}{r_i \Delta\theta}\left(T_{i,j} - T_{i,j-1}\right), \quad (9.61)$$

in which $\frac{dT_{i,j}}{dt}$ at time $n\Delta t$ can be approximated using any time difference scheme, such as the forward scheme, $\frac{T_{i,j}^{n+1} - T_{i,j}^n}{\Delta t}$. For the other cells, shown in Fig. 9.8, we can write the following:

- Cells in which the north face is the outer boundary (yellow cells): $i = m_r$ and $j = 2, \ldots, m_\theta$. The convection boundary condition (9.57) applies there, viz.

$$-\lambda \left(\frac{\partial T}{\partial r}\right)_{r_{m_r+\frac{1}{2}}, \theta_j} = h\left(\theta_j\right)\left(T_{S_N^{m_r,j}} - T_\infty\right). \quad (9.62)$$

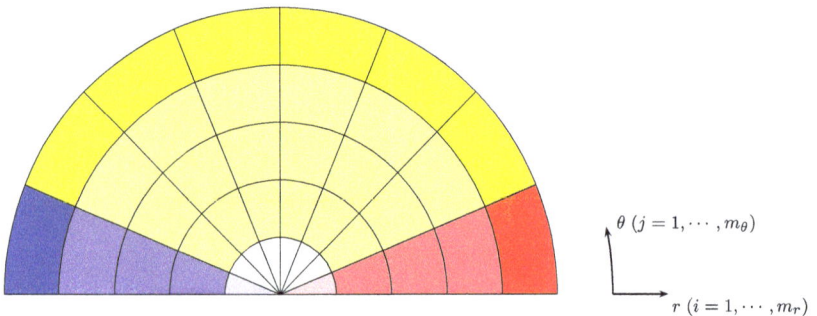

$\theta\ (j = 1, \cdots, m_\theta)$

$r\ (i = 1, \cdots, m_r)$

Figure 9.8:   Declination of the control volume according to the number of boundaries it has got.

The second-order Taylor series of T about r_{m_r} is

$$T = T_{S_N^{m_r,j}} + (r - r_{m_r})\left(\frac{\partial T}{\partial r}\right)_{r_{m_r+\frac{1}{2}},\theta_j} + O((r - r_{m_r})^2).$$

Introducing the boundary condition into this expansion gives

$$T = T_{S_N^{m_r,j}} - (r - r_{m_r})\frac{h\left(\theta_j\right)}{\lambda}\left(T_{S_N^{m_r,j}} - T_\infty\right) + O((r - r_{m_r})^2).$$

Applied at $r = r_{m_r}$, we get

$$T_{S_N^{m_r,j}} = \frac{T_{m_r,j} + \frac{h(\theta_j)\Delta r}{2\lambda}T_\infty}{1 + \frac{h(\theta_j)\Delta r}{2\lambda}} + O(\Delta r^2).$$

Introducing this expression into (9.62), multiplying both sides by $S_N^{m_r,j}$ and rearranging gives the expression of the flux at this face:

$$\lambda\left(\frac{\partial T}{\partial r}\right)_{r_{m_r+\frac{1}{2}},\theta_j} S_N^{m_r,j} = -\frac{h\left(\theta_j\right)}{1 + \frac{h(\theta_j)\Delta r}{2\lambda}}(T_{m_r,j} - T_\infty)S_N^{m_r,j}.$$

$$(9.63)$$

Substituting this flux into Eq. (9.61) yields the ultimate shape of the equation to be solved in this region ($i = m_r$ and $j = 2,\ldots,$ $m_\theta - 1$):

$$\rho c_p v_{m_r,j}\frac{dT_{m_r,j}}{dt} = -\frac{h\left(\theta_j\right)S_N^{m_r,j}}{1 + \frac{h(\theta_j)\Delta r}{2\lambda}}(T_{m_r,j} - T_\infty)$$

$$-\frac{\lambda S_S^{m_r,j}}{\Delta r}(T_{m_r,j} - T_{m_r-1,j})$$

$$+\frac{\lambda S_W^{m_r,j}}{r_{m_r}\Delta\theta}(T_{m_r,j+1} - T_{m_r,j})$$

$$-\frac{\lambda S_E^{m_r,j}}{r_{m_r}\Delta\theta}(T_{m_r,j} - T_{m_r,j-1}).\qquad(9.64)$$

- Cells in which the east face is along the symmetry axis (light-red cells): $i = 2,\ldots,m_r - 1$ and $j = 1$. We can check that the area of this face $S_E^{i,j}$ is null. The flux contribution of this face can be dropped from Eq. (9.61), which yields the following equation:

$$\rho c_p v_{i,j} \frac{dT_{i,j}}{dt} = \frac{\lambda S_N^{i,j}}{\Delta r}\left(T_{i+1,j} - T_{i,j}\right) - \frac{\lambda S_S^{i,j}}{\Delta r}\left(T_{i,j} - T_{i-1,j}\right)$$

$$+ \frac{\lambda S_W^{i,j}}{r_i \Delta \theta}\left(T_{i,j+1} - T_{i,j}\right). \tag{9.65}$$

- Cells in which the west face is along the symmetry axis (light-blue cells): $i = 2, \ldots, m_r - 1$ and $j = m_\theta$. We can check that the area of this face $S_W^{i,j}$ is null. The flux contribution of this face can be dropped from Eq. (9.61), which yields the following equation:

$$\rho c_p v_{i,j} \frac{dT_{i,j}}{dt} = \frac{\lambda S_N^{i,j}}{\Delta r}\left(T_{i+1,j} - T_{i,j}\right) - \frac{\lambda S_S^{i,j}}{\Delta r}\left(T_{i,j} - T_{i-1,j}\right)$$

$$- \frac{\lambda S_E^{i,j}}{r_i \Delta \theta}\left(T_{i,j} - T_{i,j-1}\right). \tag{9.66}$$

- Cells in which the south face is at the origin (very-light-yellow cells): $i = 1$ and $j = 2, \ldots, m_\theta - 1$. We can check that the area of this face $S_S^{i,j}$ is null. The flux contribution of this face can be dropped from Eq. (9.61), which yields the following equation:

$$\rho c_p v_{i,j} \frac{dT_{i,j}}{dt} = \frac{\lambda S_N^{i,j}}{\Delta r}\left(T_{i+1,j} - T_{i,j}\right) + \frac{\lambda S_W^{i,j}}{r_i \Delta \theta}\left(T_{i,j+1} - T_{i,j}\right)$$

$$- \frac{\lambda S_E^{i,j}}{r_i \Delta \theta}\left(T_{i,j} - T_{i,j-1}\right). \tag{9.67}$$

- Cells in which the north face is the outer boundary and the east face is along the symmetry axis (red cell): $i = m_r$ and $j = 1$. As the area $S_E^{i,j}$ of the latter is null, the flux contribution of this face can be dropped from Eq. (9.64), which yields the following equation:

$$\rho c_p v_{m_r,j} \frac{dT_{m_r,j}}{dt} = -\frac{h\left(\theta_j\right) S_N^{m_r,j}}{1 + \frac{h(\theta_j)\Delta r}{2\lambda}}\left(T_{m_r,j} - T_\infty\right)$$

$$- \frac{\lambda S_S^{m_r,j}}{\Delta r}\left(T_{m_r,j} - T_{m_r-1,j}\right)$$

$$+ \frac{\lambda S_W^{m_r,j}}{r_{m_r} \Delta \theta}\left(T_{m_r,j+1} - T_{m_r,j}\right). \tag{9.68}$$

- Cells in which the north face is the outer boundary and the west face is along the symmetry axis (blue cell): $i = m_r$ and $j = m_\theta$. As the area $S_W^{i,j}$ of the latter is null, the flux contribution of this face can be dropped from Eq. (9.64), which yields the following equation:

$$\rho c_p v_{m_r,j} \frac{dT_{m_r,j}}{dt} = -\frac{h(\theta_j) S_N^{m_r,j}}{1 + \frac{h(\theta_j)\Delta r}{2\lambda}} (T_{m_r,j} - T_\infty)$$

$$-\frac{\lambda S_S^{m_r,j}}{\Delta r} (T_{m_r,j} - T_{m_r-1,j})$$

$$-\frac{\lambda S_E^{m_r,j}}{r_{m_r}\Delta\theta} (T_{m_r,j} - T_{m_r,j-1}). \quad (9.69)$$

- Cells in which the south face is at the origin and the east face is along the symmetry axis (very-light-red cell): $i = 1$ and $j = 1$. As the area $S_E^{i,j}$ of the latter is null, the flux contribution of that face can be dropped from Eq. (9.67), which yields the following equation:

$$\rho c_p v_{i,j} \frac{dT_{i,j}}{dt} = \frac{\lambda S_N^{i,j}}{\Delta r} (T_{i+1,j} - T_{i,j}) + \frac{\lambda S_W^{i,j}}{r_i \Delta\theta} (T_{i,j+1} - T_{i,j}). \quad (9.70)$$

- Cells in which the south face is at the origin and the west face is along the symmetry axis (very-light-blue cell): $i = 1$ and $j = m_\theta$. As the area $S_W^{i,j}$ of the latter is null, the flux contribution of that face can be dropped from Eq. (9.67), which yields the following equation:

$$\rho c_p v_{i,j} \frac{dT_{i,j}}{dt} = \frac{\lambda S_N^{i,j}}{\Delta r} (T_{i+1,j} - T_{i,j}) - \frac{\lambda S_E^{i,j}}{r_i \Delta\theta} (T_{i,j} - T_{i,j-1}). \quad (9.71)$$

The set of Eqs. (9.61), (9.64), (9.65), (9.66), (9.67), (9.68), (9.69), (9.70), and (9.71) is now closed and can be implemented.

Figure 9.9 shows the temperature map, given by FVM, at different times, of a glass spherical object of radius 10^{-2} subjected to initial temperature $T_0 = 0$ and infinite temperature $T_\infty = 1$. The heat exchange coefficient is given by $h(\theta) = K \sin\theta$, where $K = 200$.

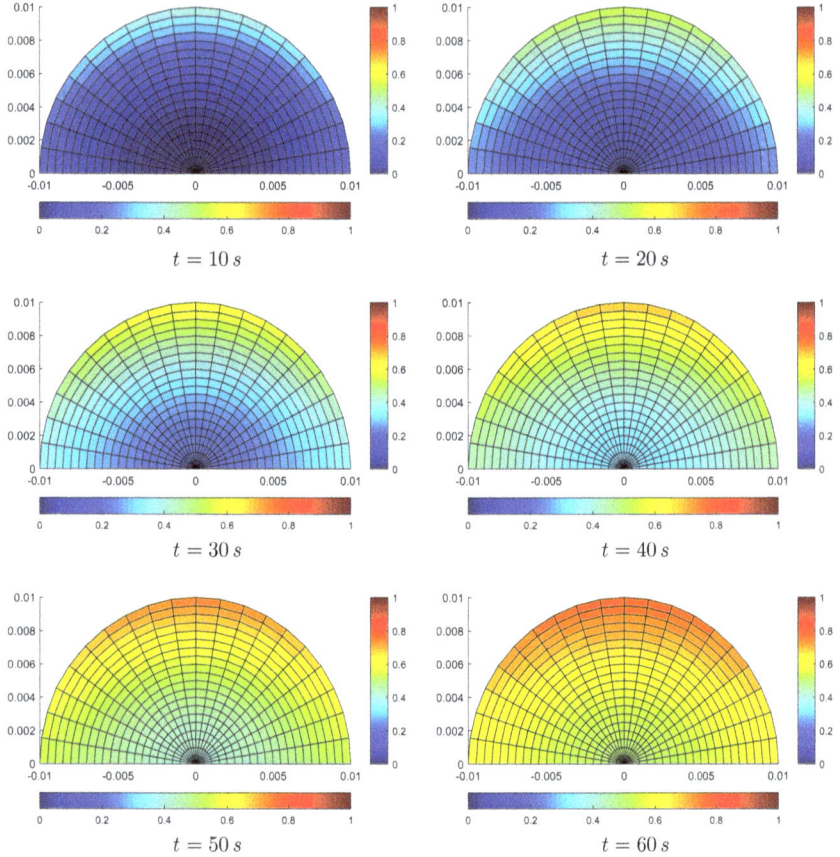

Figure 9.9:   Temperature map of the spherical body computed at different times using the FVM.

Project 1: Solve the heat diffusion in a slab governed by the equation

$$\frac{\partial T}{\partial t} = \alpha \left(\frac{\partial^2 T}{\partial x^2} + \frac{\partial^2 T}{\partial y^2} \right). \tag{9.72}$$

Initially, the slab was at room temperature ($T_i = 20°\text{C}$). Suddenly, the left-hand temperature is raised to $100°\text{C}$, while the other boundary conditions are fixed, as shown in Fig. 9.10. Compare the results obtained using FDM and FVM. Use explicit and implicit methods. Plot the temperature along the center line $T(x, H/2)$ at different time intervals.

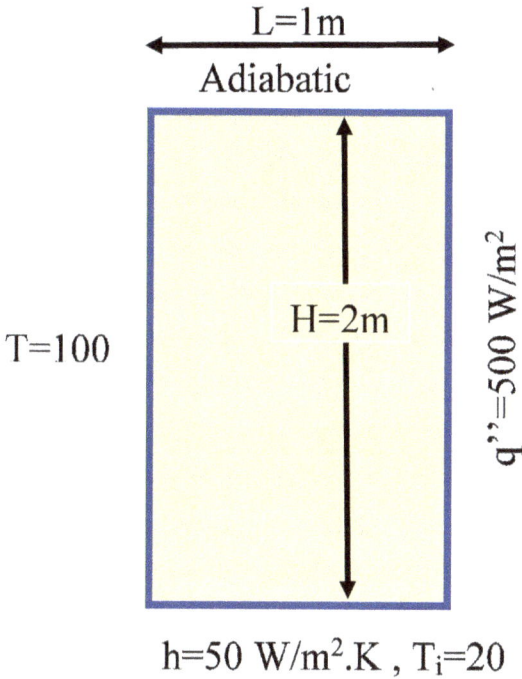

L=1m
Adiabatic

H=2m

T=100

q''=500 W/m²

h=50 W/m².K , Tᵢ=20

Figure 9.10: Slab diagram.

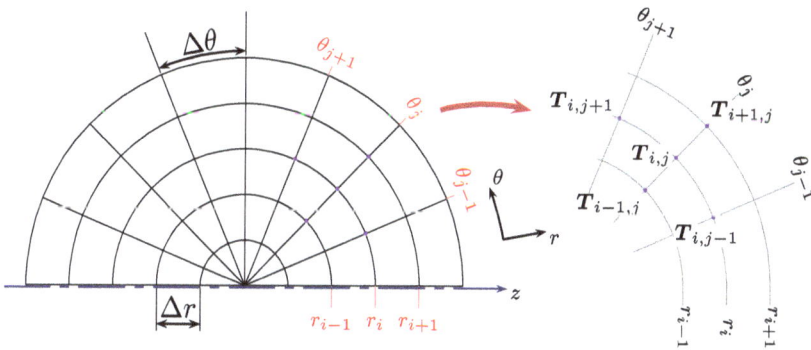

Figure 9.11: The FDM mesh (projected on the plan (xOz)) used to solve the heat transfer problem inside a spherical body.

Use $\Delta x = 0.01$ and $\Delta y = 0.02$. Select Δt to satisfy the stability conditions for explicit method. Use double time step of the explicit method for the implicit method.

Project 2: Solve the problem invoked in Section 9.7 using the finite difference method. The mesh is illustrated in Fig. 9.11.

Hints: Expressed in spherical coordinates, Laplace operator applied to T, which appears in Eq. (9.55), reads

$$\nabla \cdot \nabla T = \nabla^2 T = \frac{1}{r^2} \frac{\partial}{\partial r} \left(r^2 \frac{\partial T}{\partial r} \right) + \frac{1}{r^2 \sin \theta} \frac{\partial}{\partial \theta} \left(\sin \theta \frac{\partial T}{\partial \theta} \right)$$

$$+ \frac{1}{r^2 \sin^2 \theta} \frac{\partial^2 T}{\partial \varphi^2}.$$

Use symmetry with respect to φ to cancel irrelevant terms. Because of that very same symmetry, we have $\frac{\partial T}{\partial \theta} = 0$ along the z-axis, i.e., when $\theta = 0$ or $\theta = \pi$. So, to remove the indetermination along this axis, apply L'Hospital's rule of limit with the term $\left(\frac{\cos \theta}{r^2 \sin \theta} \frac{\partial T}{\partial \theta} \right)$, when $\theta \to 0$ or $\theta \to \pi$. Also, at the origin $r = 0$, it can be proven that Laplace operator degenerates into $3 \left. \frac{\partial^2 T}{\partial r^2} \right|_{r=0}$. (Beware of the factor 3.)

Chapter 10

Laplace and Poisson Equations (Elliptic Equations)

10.1 Introduction

Equilibrium problems in physics and engineering are governed by elliptic equations, e.g., potential theory, heat conduction, electrostatics, and magnetostatics. An example of elliptic equation is Poisson equation that can be written in two-dimensional Cartesian coordinate system as

$$\frac{\partial^2 f}{\partial x^2} + \frac{\partial^2 f}{\partial y^2} = F(x, y). \tag{10.1}$$

If the force function $F(x, y)$ is zero, the equation is called Laplace equation, the simplest of all elliptic equations. The methodology of solution is the same. To close the solution, two boundary conditions are required in $x-$ and $y-$ directions, i.e., a total of four boundary conditions. In fact, the conditions at the boundary determine the solution. In other words, only the boundary conditions decide the domain solution for Laplace equation. For Poisson equation, the force (source or sink) term also contributes to the solution. Theoretically, the information should propagate from the boundaries into the domain instantly (with infinite speed). Practically, most numerical solutions are iterative in nature and, consequently, the speed of information is finite. The best algorithm is the one that propagates the information from the boundary into the domain as fast as possible.

10.2   Finite Difference Method

Let us consider Laplace equation by setting the force term in
Eq. (10.1) to zero ($F(x, y) = 0$). The finite difference (FD) approximation for this equation is

$$\frac{f_{i+1,j} - 2f_{i,j} + f_{-1,j}}{\Delta x^2} + \frac{f_{i,j+1} - 2f_{i,j} + f_{i,j-1}}{\Delta y^2} = 0. \qquad (10.2)$$

The above approximation is second-order accurate in space
($O(\Delta x^2, \Delta y^2)$). It can be reformulated as

$$f_{i+1,j} + f_{i-1,j} + \beta f_{i,j+1} + \beta f_{i,j-1} - 2(1 + \beta)f_{i,j} = 0, \qquad (10.3)$$

where $\beta = (\Delta x/\Delta y)^2$. These equations form a system of algebraic
equations with penta-diagonal banded matrix. It is possible to use
higher-order accurate schemes. However, extra points (nodes) need
to be involved making: (i) the matrix even less sparse and (ii) the
way to deal with the boundaries more complicated.

10.2.1   Worked example

Solve the following equation for the given boundary conditions of the
domain, see Fig. 10.1:

$$\frac{\partial^2 \phi}{\partial x^2} + \frac{\partial^2 \phi}{\partial y^2} = 2. \qquad (10.4)$$

Discretization for the FDM: The domain is discretized as illustrated in Fig. 10.2.

Step 1: Discretize Eq. (10.4) for an interior node (i, j) (a node that
has exactly four neighbors):

$$\left.\frac{\partial^2 \phi}{\partial x^2}\right|_{i,j} = \frac{\phi_{i+1,j} - 2\phi_{i,j} + \phi_{i-1,j}}{\Delta x^2} = \frac{\phi_E - 2\phi_P + \phi_W}{\Delta x^2},$$

$$\left.\frac{\partial^2 \phi}{\partial y^2}\right|_{i,j} = \frac{\phi_{i,j+1} - 2\phi_{i,j} + \phi_{i,j-1}}{\Delta y^2} = \frac{\phi_N - 2\phi_P + \phi_S}{\Delta y^2}.$$

Plug the above into Eq. (10.4):

$$\frac{\phi_E - 2\phi_P + \phi_W}{\Delta x^2} + \frac{\phi_N - 2\phi_P + \phi_S}{\Delta y^2} = 2,$$

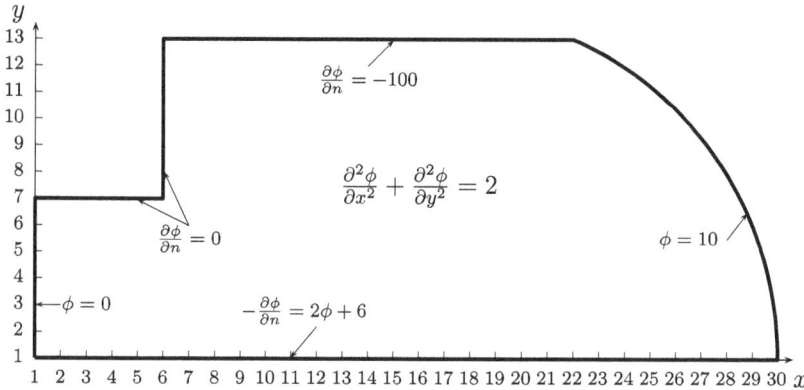

Figure 10.1: Sketch of the two-dimensional Poisson problem considered. The curved boundary is an arc of the center located at indices $(17, 1)$ and radius 13 units of length. (Note: The node $(1, 1)$ is located at the origin $(0, 0)$.)

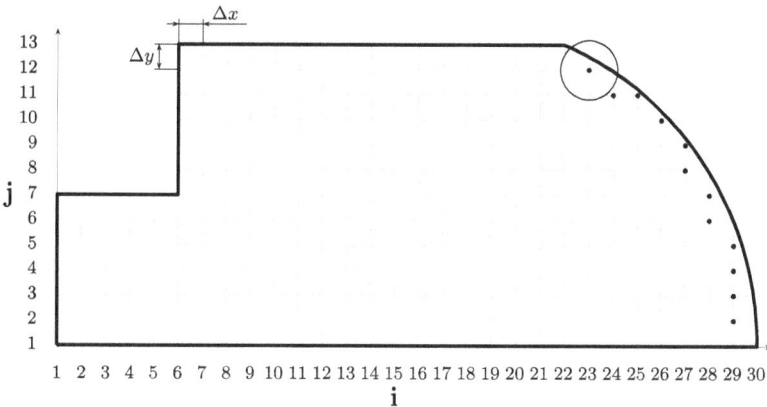

Figure 10.2: Discretization of the domain.

which can be rearranged as

$$\underbrace{2\left(\frac{1}{\Delta x^2} + \frac{1}{\Delta y^2}\right)}_{a_P}\phi_P = \underbrace{\left(\frac{1}{\Delta x^2}\right)}_{a_W}\phi_W + \underbrace{\left(\frac{1}{\Delta x^2}\right)}_{a_E}\phi_E$$

$$+ \underbrace{\left(\frac{1}{\Delta y^2}\right)}_{a_S}\phi_S + \cdot \underbrace{\left(\frac{1}{\Delta y^2}\right)}_{a_N}\phi_N - 2.$$

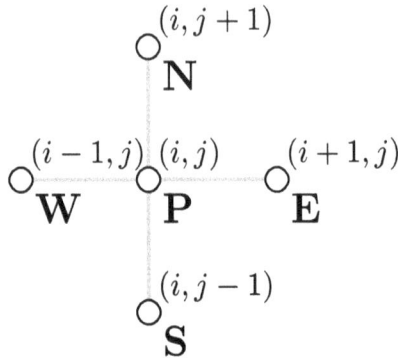

Figure 10.3:   Discretization of interior node.

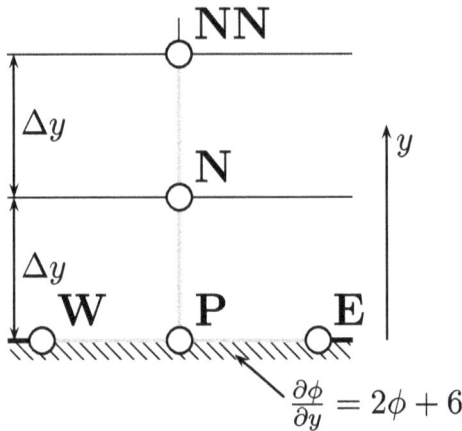

Figure 10.4:   Discretization of bottom nodes.

Figures 10.3 and 10.4, illustrated typical nodes at the interior domain and at the bottom boundary of the domain, respectively. Hence,

$$a_P \phi_P = a_W \phi_W + a_E \phi_E + a_S \phi_S + a_N \phi_N - 2. \tag{10.5}$$

Step 2: On the south boundary, the nodes considered are $(i, 1)$, where $i = 2, 3, \ldots, 29$. Let us write $\phi = A + By + Cy^2$. We can determine A, B, and C based on the following three conditions:

At $y = 0$, $\frac{\partial \phi}{\partial y} = 2\phi + 6$ (note here that $\frac{\partial}{\partial n} = -\frac{\partial}{\partial y}$).
At $y = \Delta y$, $\phi = \phi_N$.
At $y = 2\Delta y$, $\phi = \phi_{NN}$.

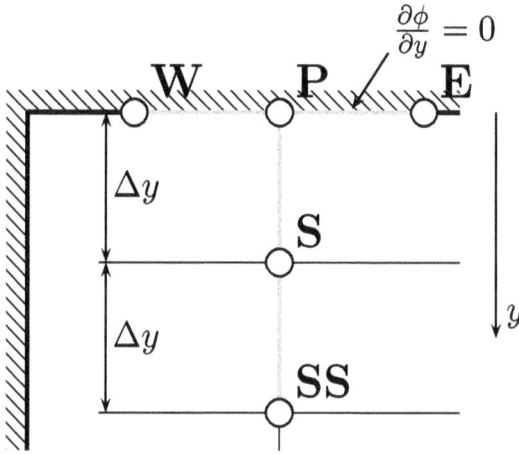

Figure 10.5: Discretization of middle corner nodes, horizontal wall.

Figure 10.5 illustrates typical nodes at the left side, upper corner of the domain. By solving the above three equations, one will get the following:

$$\phi = \underbrace{\frac{4\phi_N - \phi_{NN}}{3} - \frac{2(2\phi_P + 6)\Delta y}{3}}_{A} + \underbrace{(2\phi_P + 6)\,y}_{B}$$

$$+ \underbrace{\left(\frac{\phi_{NN} - \phi_N}{3\Delta y^2} - \frac{2\phi_P + 6}{3\Delta y} \right) y^2}_{C}.$$

As we need only the value for $y = 0$, we can write that $\phi_P = A$. And from here, one can write the equation applicable to all nodes $(i, 1)$:

$$\phi_P = \frac{4\phi_N - \phi_{NN}}{3 + 4\Delta y} - \frac{12\Delta y}{3 + 4\Delta y}. \tag{10.6}$$

Step 3: Proceed in a similar manner for all nodes $(i, 7)$, where $i = 2, \ldots, 5$.

Let us write $\phi = A + By + Cy^2$. We can determine A, B, and C based on the following three conditions:

At $y = 0$, $\frac{\partial \phi}{\partial y} = 0$.
At $y = \Delta y$, $\phi = \phi_S$.
At $y = 2\Delta y$, $\phi = \phi_S$.

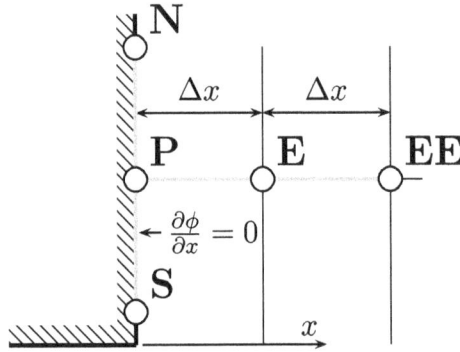

Figure 10.6:   Discretization of middle corner nodes, vertical wall.

Figure 10.6 illustrates typical nodes at the left side wall of the bottom corner of the wall. By solving the above three equations, one will get the following:

$$\phi = \underbrace{\frac{4\phi_S - \phi_{SS}}{3}}_{A} + \underbrace{0}_{B} + \underbrace{\frac{\phi_{SS} - \phi_S}{3\Delta y^2}}_{C} y^2.$$

As we need only the value for $y = 0$, we can write that $\phi_P = A$. And from here, one can write the equation applicable to all nodes $(i, 7)$, $i = 2, \ldots, 5$:

$$\phi_P = \frac{4\phi_S - \phi_{SS}}{3}. \tag{10.7}$$

Step 4: Proceed in a similar manner for all nodes $(i, 13)$, where $i = 7, \ldots, 21$. Consequently, for all these nodes, one can write the following equation:

$$\phi_P = \frac{4\phi_E - \phi_{EE}}{3}. \tag{10.8}$$

Step 5: Proceed in a similar manner for all nodes $(i, 10)$, where $i = 7, \ldots, 21$. Let us write $\phi = A + By + Cy^2$. We can determine A, B, and C based on the following three conditions:

At $y = 0$, $\frac{\partial \phi}{\partial y} = 100$ (note here that $\frac{\partial}{\partial n} = -\frac{\partial}{\partial y}$).
At $y = \Delta y$, $\phi = \phi_S$.
At $y = 2\Delta y$, $\phi = \phi_S$.

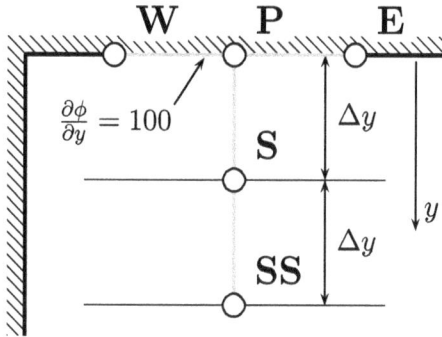

Figure 10.7: Discretization of top wall nodes.

Figure 10.7 illustrates the nodes at the upper horizontal boundary of the domain. By solving the above three equations, one will get the following:

$$\phi = \underbrace{\frac{4\phi_S - \phi_{SS}}{3} - \frac{200}{3}\Delta y}_{A} + \underbrace{100}_{B}\,y + \underbrace{\frac{\phi_{SS} - \phi_S}{3\Delta y^2} - \frac{100}{3\Delta y}\,y^2}_{C}.$$

As we need only the value for $y = 0$, we can write that $\phi_P = A$. And from here, one can write the equation applicable to all nodes $(i, 13)$, $i = 7, \ldots, 21$:

$$\phi_P = \frac{4\phi_S - \phi_{SS}}{3} - \frac{200}{3}\Delta y. \tag{10.9}$$

Step 6: All nodes located near the curved boundary (filled circles in Fig. 10.3) need a special treatment. Let us consider the case of such a node, presented in Fig. 10.8. The distances from this node in the horizontal and vertical directions toward the curve (where the value is known, $\phi = 10$) are $L_{i,j}$ and $H_{i,j}$, respectively. These distances are considered to be known from both the geometry and the mesh set.

Let us develop a relation for the second-order derivative of a function using a non-uniform grid, which is the case for our curved boundary. We use Taylor series expansion and the domain is presented in Fig. 10.9:

$$\phi(x, y) = \phi(x_i, y) + \left.\frac{\partial \phi}{\partial x}\right|_{x_i, y} (x - x_i) + \left.\frac{\partial^2 \phi}{\partial x^2}\right|_{x_i, y} \frac{(x - x_i)^2}{2}.$$

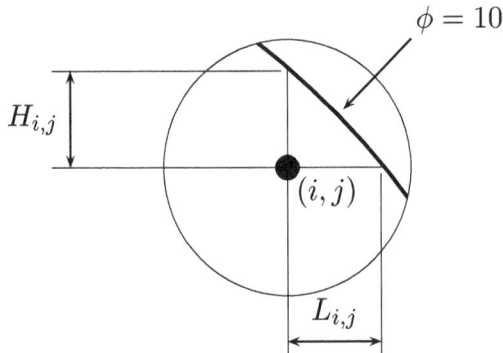

Figure 10.8:   Closeup of a node next the curved wall.

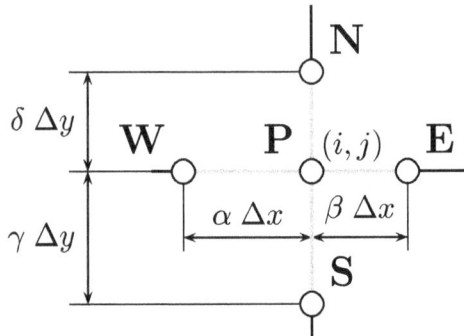

Figure 10.9:   Curved wall nodes.

From the above, we can write that

$$\phi_{i+1,j} = \phi_{i,j} + \left.\frac{\partial \phi}{\partial x}\right|_{i,j} \beta\,\Delta x + \left.\frac{\partial^2 \phi}{\partial x^2}\right|_{i,j} \frac{(\beta\,\Delta x)^2}{2},$$

$$\phi_{i-1,j} = \phi_{i,j} - \left.\frac{\partial \phi}{\partial x}\right|_{i,j} \alpha\,\Delta x + \left.\frac{\partial^2 \phi}{\partial x^2}\right|_{i,j} \frac{(\alpha\,\Delta x)^2}{2}.$$

Dividing the first equation by $\beta\,\Delta x$, the second by $\alpha\,\Delta x$, and then summing them side by side yields

$$\frac{\phi_{i+1,j}}{\beta\,\Delta x} + \frac{\phi_{i-1,j}}{\alpha\,\Delta x} = \phi_{i,j}\left(\frac{\alpha+\beta}{\alpha\beta}\right)\frac{1}{\Delta x} + \left.\frac{\partial^2 \phi}{\partial x^2}\right|_{i,j} \frac{(\alpha+\beta)\Delta x}{2}.$$

From here, we can write that

$$\frac{\partial^2 \phi}{\partial x^2}\Big|_{i,j} = \left(\frac{2}{\alpha(\alpha+\beta)\Delta x^2}\right)\phi_{i-1,j} - \left(\frac{2}{\alpha\beta\Delta x}\right)\phi_{i,j}$$
$$+ \left(\frac{2}{\beta(\alpha+\beta)\Delta x^2}\right)\phi_{i+1,j}$$

or

$$\frac{\partial^2 \phi}{\partial x^2}\Big|_{P} = \left(\frac{2}{\alpha(\alpha+\beta)\Delta x^2}\right)\phi_W - \left(\frac{2}{\alpha\beta\,\Delta x}\right)\phi_P$$
$$+ \left(\frac{2}{\beta(\alpha+\beta)\Delta x^2}\right)\phi_E. \qquad (10.10)$$

Analogously, it can be shown that

$$\frac{\partial^2 \phi}{\partial y^2}\Big|_{P} = \left(\frac{2}{\gamma(\gamma+\delta\Delta x^2)}\right)\phi_W - \left(\frac{2}{\gamma\delta\,\Delta x}\right)\phi_P + \left(\frac{2}{\delta(\gamma+\beta)\Delta x^2}\right)\phi_E.$$
$$(10.11)$$

Now, we can discretize Eq. (10.4) for such points where we have a variable grid. Plugging (10.10) and (10.11) into (10.4) gives

$$\frac{2}{(\alpha+\beta)\Delta x^2}\left[\left(\frac{1}{\alpha}\right)\phi_W - \left(\frac{\alpha+\beta}{\alpha\beta}\right)\phi_P + \left(\frac{1}{\beta}\right)\phi_E\right]$$
$$+ \frac{2}{(\gamma+\delta)\Delta x^2}\left[\left(\frac{1}{\gamma}\right)\phi_W - \left(\frac{\gamma+\delta}{\gamma\delta}\right)\phi_P + \left(\frac{1}{\delta}\right)\phi_E\right] = 2.$$

Hence, the equation for such points is

$$\underbrace{\left(\frac{1}{\alpha_{i,j}\beta_{i,j}\Delta x^2} + \frac{1}{\gamma_{i,j}\delta_{i,j}\Delta y^2}\right)}_{AP_{i,j}}\phi_P$$

$$= \underbrace{\frac{1}{\alpha_{i,j}(\alpha_{i,j}+\beta_{i,j})\Delta x^2}}_{AW_{i,j}}\phi_W + \underbrace{\frac{1}{\beta_{i,j}(\alpha_{i,j}+\beta_{i,j})\Delta x^2}}_{AE_{i,j}}\phi_E$$

$$+ \underbrace{\frac{1}{\gamma_{i,j}(\gamma_{i,j}+\delta_{i,j})\Delta y^2}}_{AS_{i,j}}\phi_S + \underbrace{\frac{1}{\delta_{i,j}(\gamma_{i,j}+\delta_{i,j})\Delta y^2}}_{AB_{i,j}}\phi_N - 1.$$

In a simpler form,

$$AP_{i,j}\,\phi_P = AW_{i,j}\,\phi_W + AE_{i,j}\,\phi_E + AS_{i,j}\,\phi_S + AN_{i,j}\,\phi_N - 1. \qquad (10.12)$$

Finally, we have a system of linear equations, which can be solved using one of the methods presented in Chapter 2.

10.3   Finite Volume Method

As discussed previously, finite volume (FV) method is easy to apply. The governing equations need to be multiplied by the finite volume of the domain and integrated. The two-dimensional Poisson equation in Cartesian coordinate system is

$$\frac{\partial^2 f}{\partial x^2} + \frac{\partial^2 f}{\partial y^2} = S(x, y, f), \qquad (10.13)$$

where the finite volume is $dx\,dy$. Let us work this equation term by term.

The first term reads

$$\int_s^n \int_w^e \frac{\partial}{\partial x}\left(\frac{\partial f}{\partial x}\right) dx\,dy = \left[\left.\frac{\partial f}{\partial x}\right|_e - \left.\frac{\partial f}{\partial x}\right|_w\right]\Delta y, \qquad (10.14)$$

which ends up as

$$\left[\frac{f_E - f_P}{\delta x_e} - \frac{f_P - f_W}{\delta x_w}\right]\Delta y = a_E f_E - a_{Px} f_P + a_W f_W, \qquad (10.15)$$

where a_E, a_W, and a_{Px} are $\Delta y/\delta x_e$, $\Delta y/\delta x_w$, and $(a_E + a_W)$, respectively. Similarly, the second term will end up as

$$\left[\frac{f_N - f_P}{\delta y_n} - \frac{f_P - f_S}{\delta y_s}\right]\Delta x = a_N f_N - a_{Py} f_P + a_S f_S, \qquad (10.16)$$

where a_N, a_S, and a_{Py} are $\Delta x/\delta y_n$, $\Delta x/\delta y_s$, and $(a_N + a_S)$, respectively.

The source term $S(x, y, f) = S(x_P, y_P, f_P)\Delta x \Delta y$. By adding all the terms, we end up with the following algebraic equation of each control volume:

$$a_p f_P = a_E f_E + a_W f_W + a_N f_N + a_S f_S - S\Delta x \Delta y, \qquad (10.17)$$

where $a_P = a_E + a_W + a_N + a_S$. For the notations, see Fig. 7.2 in Chapter 7 showing the diagram of a finite volume. Note that a_P is the diagonal element of the matrix, which is equal to the sum of all the elements on the same raw.

10.4   Methods of Solution

System of algebraic equations retrieved from approximations of the elliptic equations is diagonally dominant. Hence, any method (Gauss–Seidel, SOR, etc.) discussed in Chapter 2 converges. However, fast convergence is very important. Elliptic equation solution is dictated by the boundary. In other words, the boundary information (zero time) affects instantly the domain solution. Since we are using iterative methods, there is no way that the information instantly propagates into the domain. Most of iterative methods discussed earlier propagate pointwise. Therefore, their convergence is slow.

The alternating direct implicit (ADI) method explained in Chapter 9 can equally be applied to Laplace and Poisson equations. The method is much faster than pointwise iterative methods because the method introduces the information of the boundaries instantly.

As mentioned in Chapter 9, the main idea behind the ADI method is to convert the resulting matrix into a tri-diagonal matrix by moving all elements to the right side of the matrix equation, except the diagonal, upper diagonal, and lower diagonal elements.

Consider the following equation:

$$\frac{\partial^2 f}{\partial x^2} + \frac{\partial^2 f}{\partial y^2} = S. \tag{10.18}$$

Using FDM and considering implicit in x-direction and explicit in y-direction yields

$$\frac{f_{i+1,j}^{n+1} - 2f_{i,j}^{n+1} + f_{i-1,j}^{n+1}}{\Delta x^2} = -\frac{f_{i,j+1}^n - 2f_{i,j}^{n+1} + f_{i,j-1}^n}{\Delta y^2} + S_{i,j}^n. \tag{10.19}$$

Note that superscripts n and $n+1$ represent the old and updated values through the iterations process, respectively. They indicate in no mean time for the problem is intrinsically time-independent, i.e., steady. The above equation can be arranged as

$$a_w f_{i-1,j}^{n+1} - 2a_p f_{i,j}^{n+1} + a_e f_{i+1,j}^{n+1} = -\frac{f_{i,j+1}^n + f_{i,j-1}^n}{\Delta y^2} + S_{i,j}^n, \quad (10.20)$$

where $a_w = a_e = 1/\Delta x^2$ and $a_p = a_w + a_e$. The iterations are carried out until the error norm satisfies the convergence criterion within a prescribed tolerance.

The right-hand side of the equation consists of guessed values (known). As soon as those values are available, they replace the guessed ones. The sweeping processes of the domain alternates between $x-$ and $y-$ directions, as discussed in Chapter 9.

Another idea to accelerate the convergence is to use a method called **chain method**. Instead of applying tri-diagonal matrix line by line, the method converts all of the domain into one chain and applies Thomas algorithm to the domain at once. The key idea is to map the domain matrix into one tri-diagonal matrix rather than to a few tri-diagonal matrices as in ADI method, see Fig. 10.10.

Hence, rather than labeling the dependent variables by (i, j), where $i = 1, \ldots, m$ and $j = 1, \ldots, n$, we label them as one (long) vector:

$$(1, 2, 3, \ldots, n, n+1, \ldots, 2n, 2n+1, \ldots, (m-1)n, \ldots,$$
$$(m-1)n+1, \ldots, mn),$$

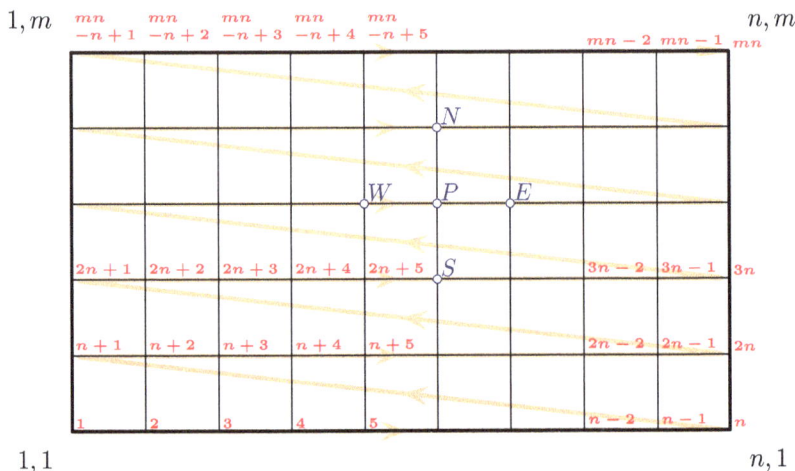

Figure 10.10:   Illustration of the chain method.

with a unique index l given by $l = (i-1)n + j$ and $l = 1, \ldots, mn$.
Variants of Thomas algorithm can be applied.

10.5 Problems

Problem 1: Flow in a micro-tube can be modeled as Stoke's flow,
which is the momentum equation with negligible advection terms
(nonlinear terms):

$$0 = -\frac{dP}{dz} + \frac{1}{\text{Re}}\left[\frac{\partial^2 U}{\partial z^2} + \frac{1}{r}\frac{\partial}{\partial r}\left(r\frac{\partial U}{\partial r}\right)\right], \qquad (10.21)$$

where Re is the Reynolds number.

Use FD and FV methods to predict the velocity profile at $z = 5$
and $z = 10$; use 50×100 nodes or control volumes along $z-$ and
$r-$ directions, respectively. Set Re $= 2.0$ and $dP/dz = -3/2$.

Hint: The finite volume is $2\pi r\, dr\, dz$.

Problem 2: A laser beam is applied to a surface for a period of
time until it reached steady-state conditions. The problem can be
modeled as shown in Fig. 10.11. The laser beam has maximum
power of 1000 W/m^2 and is applied at the middle of the domain.
The domain has the following thermophysical properties: thermal

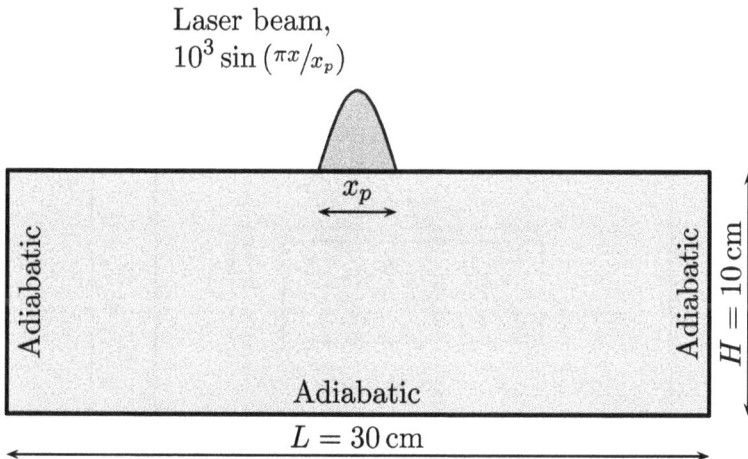

Figure 10.11: Illustration of the laser beam problem.

conductivity: $k = 0.6$ W/m K, density: $\rho = 1000$ kg/m^3, and specific heat: $c = 4180$ J/kg K. The thermal diffusivity is $\alpha = \frac{k}{\rho c}$. All the boundaries are adiabatic except the upper layer, which is exposed to the laser beam, with $x_p = 2$ mm. The other part of the upper boundary is exposed to the convection boundary conditions as

$$-k\frac{\partial T}{\partial n}\bigg|_s = h(T_s - T_a). \tag{10.22}$$

The notation s refers to the surface, and n is the normal direction to the surface. The heat transfer coefficient h is 10 W/m^{2}K. Use FV method to solve the governing equations, that is, Laplace equation:

$$\frac{\partial^2 T}{\partial x^2} + \frac{\partial^2 T}{\partial y^2} = 0. \tag{10.23}$$

Plot the temperature contours. Use 90×30 volumes in x and y directions, respectively. Repeat the solution for isothermal bottom boundary condition, $T = 0$. Note that you can utilize the symmetry condition of the problem, hence the computation cost can be reduced by half.

Problem 3: A two-dimensional block with thermal conductivity of $k_1 = 2$ W/m K is fitted with another block with a thermal conductivity of $k_2 = 12$ W/m K. The second block is centered inside the main block. The dimensions of both blocks are given in Fig. 10.12.

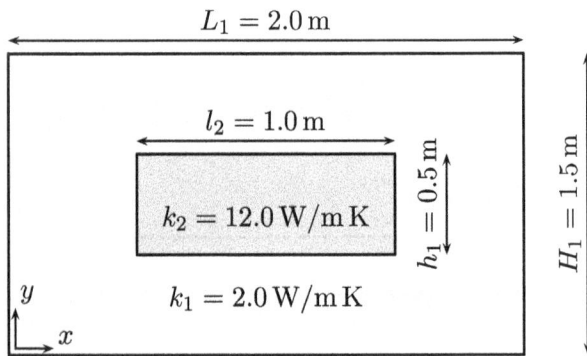

Figure 10.12:   Illustration of the nested blocks problem.

The following boundary conditions are imposed on the main block:

Left: $T = 100°$C; Right: $T = 20°$C; Bottom: adiabatic, $q = 0$ W/m^2; Top: flux $q = 100$ W/m^2. The governing equation is Laplace equation. Solve the problem using FV and FD methods for a grid of 40×30 in x and y directions, respectively. Plot the temperature distributions versus x for $y = 0.75$ m and versus y for $x = 1.0$ m. Compare both solution methods.

Chapter 11

Advection and Advection–Diffusion Equations

Advection equation is a simple wave equation, i.e., a hyperbolic equation. In a one-dimensional problem, linear advection equation can be written as

$$\frac{\partial f}{\partial t} + u\frac{\partial f}{\partial x} = 0, \tag{11.1}$$

where u is the physical velocity (characteristic velocity). If u is constant, the information propagates along a characteristic line (or curve, in general), $x - ut$. The function f keeps its properties as it moves from one location to another without distortion or diffusion. Many schemes were developed to solve such equations.

11.1 Explicit Methods

Explicit, forward in time and forward in space (FTFS) scheme applied to advection equation reads

$$\frac{f_i^{n+1} - f_i^n}{\Delta t} = -u\frac{f_{i+1}^n - f_i^n}{\Delta x}. \tag{11.2}$$

Rearranging, the equation yields

$$f_i^{n+1} = \left(1 + \frac{u\Delta t}{\Delta x}\right)f_i^n - \frac{u\Delta t}{\Delta x}f_{i+1}^n, \tag{11.3}$$

221

which is unstable if $u > 0$ because it violates the physics of the problem, for the information should propagate in positive $x-$ direction and not in negative $x-$ direction as suggested by the contribution of f_{i+1}^n. Let us check the instability more formally. von Neumann stability analysis of this equation reads

$$MG^{n+1}e^{Iki\Delta x} = MG^n e^{Iki\Delta x}(1+C) - CMG^n e^{Ik(i+1)\Delta x}, \quad (11.4)$$

where $C = \frac{u\Delta t}{\Delta x}$ is a strictly positive number, which is none other than Courant number introduced in Chapter 8. If we define $n_s = \frac{\Delta x}{\Delta t}$, which is the numerical (or node) speed, then $C = \frac{u}{n_s}$ and C may be interpreted as the ratio of the physical speed to the numerical speed.

By dividing both sides by $MG^n e^{Iki\Delta x}$, setting $k\Delta x = \theta$, recalling Euler's formula $(\exp(I\,\theta) = \cos\theta + I\sin\theta)$ and rearranging, we get

$$G = 1 + C(1 - \cos\theta) - I\sin\theta. \quad (11.5)$$

The gain G here is complex, so the stability condition reads $\forall\theta \in [0;\ 2\pi): |G|^2 \leqslant 1$, *viz.*,

$$\forall\theta \in [0;\ 2\pi) : 1 + 2C(1+C)(1 - \cos\theta). \quad (11.6)$$

Obviously, $|G|^2 > 1$ for all $\theta \in (0;\ 2\pi)$ regardless of the value of $C < 0$ and the scheme is unconditionally unstable indeed.

However, if $u < 0$, the scheme is conditionally stable (left as an exercise), where the information propagates in the appropriate direction. In that case, the stability condition is that $C = \frac{|u|\Delta t}{\Delta x}$ is less than or equal to one. For the explicit scheme to be stable, the physical speed cannot be larger than the numerical speed.

11.1.1   Lax–Wendroff method

Let us examine the explicit, forward in time and central difference in space (FTCS) scheme:

$$\frac{f_i^{n+1} - f_i^n}{\Delta t} = -u\frac{f_{i+1}^n - f_{i-1}^n}{2\Delta x}. \quad (11.7)$$

The above equation can be written as

$$f_i^{n+1} = f_i^n - u\frac{\Delta t}{2\Delta x}[f_{i+1}^n - f_{i-1}^n], \quad (11.8)$$

which is unconditionally unstable as the expression of the gain reads

$$G = 1 - I\,C\sin\theta. \tag{11.9}$$

It does not matter whether u (and so C) is positive or negative, $|G| > 1$ for all $\theta \in [0;\,2\pi)$ (except for $\theta = \frac{\pi}{2}$, $\frac{3\pi}{2}$, where $|G| = 1$, but it is of no help).

However, the above equation can be modified by using Taylor series expansion for terms f_i^{n+1}, f_{i+1}^n, and f_{i-1}^n. Using Taylor series expansion yields (index i is omitted in derivatives for brevity)

$$
\begin{aligned}
f_i^n + \frac{\partial f}{\partial t}\Delta t + \frac{\Delta t^2}{2}\frac{\partial^2 f}{\partial t^2} + \cdots \\
= f_i^n - \frac{u\Delta t}{2\Delta x}\left[f_i^n + \frac{\partial f}{\partial x}\Delta x + \frac{\Delta x^2}{2}\frac{\partial^2 f}{\partial x^2} + \cdots \right.\\
\left. - \left(f_i^n - \frac{\partial f}{\partial x}\Delta x + \frac{\Delta x^2}{2}\frac{\partial^2 f}{\partial x^2} - \cdots \right) \right],
\end{aligned}
\tag{11.10}
$$

i.e.,

$$\frac{\partial f}{\partial t} = -u\frac{\partial f}{\partial x} - \frac{\Delta t^2}{2}\frac{\partial^2 f}{\partial t^2} + \cdots. \tag{11.11}$$

The last leading term is an extra term added to the original equation (11.1). Hence, if it is artificially added to the original equation, the analysis leads to a second-order accurate scheme in space and in time. Moreover, the scheme is conditionally stable. The second derivative in time can be expressed using Eq. (11.1) as

$$\frac{\partial}{\partial t}\left(\frac{\partial f}{\partial t}\right) = -u\frac{\partial}{\partial t}\left(\frac{\partial f}{\partial x}\right) = -u\frac{\partial}{\partial x}\left(\frac{\partial f}{\partial t}\right). \tag{11.12}$$

Hence,

$$\frac{\partial^2 f}{\partial t^2} = u^2\frac{\partial^2 f}{\partial x^2}. \tag{11.13}$$

The final finite difference scheme is

$$f_i^{n+1} = f_i^n - \frac{u\Delta t}{2\Delta x}\left(f_{i+1}^n - f_{i-1}^n\right) + \frac{u^2\Delta t^2}{2\Delta x^2}\left(f_{i+1}^n - 2f_i^n + f_{i-1}^n\right). \tag{11.14}$$

This scheme is called Lax–Wendroff scheme. Rearranging the equation gives

$$f_i^{n+1} = f_i^n (1 - C^2) + f_{i+1}^n \frac{C}{2}(C - 1) + f_{i-1}^n \frac{C}{2}(C + 1), \qquad (11.15)$$

where, as before, $C = \frac{u\Delta t}{\Delta x}$. The scheme is stable for $C \leqslant 1$. (The details of the stability analysis are left as an exercise).

11.1.2   Upwind scheme

The upwind scheme, taking the sign of the advection velocity u into account, reads

$$\frac{f_i^{n+1} - f_i^n}{\Delta t} = -u \frac{f_i^n - f_{i-1}^n}{\Delta x} \quad \text{if } u > 0,$$

and   (11.16)

$$\frac{f_i^{n+1} - f_i^n}{\Delta t} = -u \frac{f_{i+1}^n - f_i^n}{\Delta x} \quad \text{if } u < 0.$$

The upwind method is stable for $C \leqslant 1$. In fact, the exact solution can be obtained if $C = 1$. However, the scheme is first-order accurate in time and in space, i.e., $O(\Delta t, \Delta x)$.

11.1.3   Lax method

As mentioned earlier, the FTCS is unconditionally unstable. A simple modification to the scheme will make the scheme conditionally stable, i.e., the scheme will be stable for $C \leqslant 1$. The scheme is called Lax method, which is to replace f_i^n by the average value:

$$\frac{f_i^{n+1} - \left(\frac{f_{i+1}^n + f_{i-1}^n}{2} \right)}{\Delta t} = -u \frac{f_{i+1}^n - f_{i-1}^n}{2\Delta x}. \qquad (11.17)$$

Rearranging gives,

$$f_i^{n+1} = \frac{1}{2}(f_{i+1}^n + f_{i-1}^n) - \frac{u\Delta t}{2\Delta x}(f_{i+1}^n - f_{i-1}^n) \qquad (11.18)$$

or in a more convenient form

$$f_i^{n+1} = 0.5 \left((1 - C)f_{i+1}^n + (1 + C)f_{i-1}^n \right). \qquad (11.19)$$

11.1.4   Leapfrog method

Central difference is used for both time and space, CTCS:

$$\frac{f_i^{n+1} - f_i^{n-1}}{2\Delta t} = -u\frac{f_{i+1}^n - f_{i-1}^n}{2\Delta x} \tag{11.20}$$

or

$$f_i^{n+1} = f_i^{n-1} - C(f_{i+1}^n - f_{i-1}^n). \tag{11.21}$$

As opposed to the leapfrog scheme applied to centered diffusion equation (see Section 9.2.1 of Chapter 9), the scheme presented here is conditionally stable, with $C \leqslant 1$ (details of the stability analysis are left as an exercise). It is second-order accurate in time and in space, i.e., $O(\Delta t^2, \Delta x^2)$. However, an extra time level is needed to start the solution.

11.2   Implicit Methods

If the space derivative involved in the advection equation is calculated at the updated time, then the scheme becomes implicit.

11.2.1   Implicit Euler method

Euler BTCS can be formulated as

$$\frac{f_i^{n+1} - f_i^n}{\Delta t} = -u\frac{f_{i+1}^{n+1} - f_{i-1}^{n+1}}{2\Delta x} \tag{11.22}$$

or

$$\frac{C}{2}f_{i-1}^{n+1} - f_i^{n+1} - \frac{C}{2}f_{i+1}^{n+1} = -f_i^n. \tag{11.23}$$

The scheme is first-order accurate in time and second-order accurate in space. It is unconditionally stable. The equations form a tridiagonal matrix and need to be solved by Thomas algorithm (for example). The stability analysis can be performed by von Neumann method:

$$0.5cf_i^{n+1}e^{-I\theta} - f_i^{n+1} - 0.5cf_i^{n+1}e^{I\theta} = -f_i^n. \tag{11.24}$$

Therefore, the gain $G = \frac{f_i^{n+1}}{f_i^n}$ is

$$G = \frac{1}{1 + C\left(\frac{e^{I\theta} - e^{-I\theta}}{2}\right)}. \qquad (11.25)$$

Hence,

$$|G|^2 = \frac{1}{|1 + IC\sin\theta|^2} = \frac{1}{1 + C^2\sin^2\theta} \leqslant 1. \qquad (11.26)$$

Similarly, upwind scheme can be formulated implicitly.

11.2.2   Crank–Nicolson method

The scheme is second-order accurate in time and in space and is conditionally stable. Derivation of the stability condition is left as an exercise.

The scheme is

$$\frac{f_i^{n+1} - f_i^n}{\Delta t} = -\frac{u}{2}\left[\frac{f_{i+1}^{n+1} - f_{i-1}^{n+1}}{2\Delta x} + \frac{f_{i+1}^n - f_{i-1}^n}{2\Delta x}\right]. \qquad (11.27)$$

The equations form a tri-diagonal matrix:

$$\frac{C}{4}f_{i-1}^{n+1} - f_i^{n+1} - 0.25\frac{C}{4}f_{i+1}^{n+1} = -f_i^n + \frac{C}{4}\left(f_{i+1}^n - f_{i-1}^n\right). \qquad (11.28)$$

11.3   Multi-step Methods

Difficulties may arise in solving nonlinear problems. Multi-step methods use finite difference equations at intermediary time levels. These methods are effective in solving nonlinear hyperbolic equations. These methods are also called predictor-corrector methods.

11.3.1   Richtmyer/Lax–Wendroff multi-step method

This method is second-order accurate in time and in space. The scheme is as follows:

$$\frac{f_i^{n+1/2} - \frac{1}{2}(f_{i+1}^n + f_{i-1}^n)}{\frac{1}{2}\Delta t} = -u\frac{f_{i+1}^n - f_{i-1}^n}{2\Delta x} \qquad (11.29)$$

and

$$\frac{f_i^{n+1} - f_i^n}{\Delta t} = -u\frac{f_{i+1}^{n+/12} - f_{i-1}^{n+1/2}}{2\Delta x}. \tag{11.30}$$

Those equations can be arranged as

$$f_i^{n+1/2} = \frac{1}{2}(f_{i+1}^n + f_{i-1}^n) - \frac{C}{4}\left(f_{i+1}^n - f_{i-1}^n\right) \tag{11.31}$$

and

$$f_i^{n+1} = f_i^n - \frac{C}{2}\left(f_{i+1}^{n+1/2} - f_{i-1}^{n+1/2}\right). \tag{11.32}$$

The scheme is stable for $C \leqslant 2$.

11.3.2 MacCormack method

This method is a two-step scheme, first in prediction step, use a forward difference,

$$\frac{f_i^* - f_i^n}{\Delta t} = -u\frac{f_{i+1}^n - f_i^n}{\Delta x}. \tag{11.33}$$

In correction step, use a backward difference,

$$\frac{f_i^{n+1} - 0.5(f_i^n + f_i^*)}{\Delta t/2} = -u\frac{f_i^* - f_{i-1}^*}{\Delta x}. \tag{11.34}$$

In the final form, the predictor step reads

$$f_i^* = f_i^n - C\left(f_{i+1}^n - f_i^n\right) \tag{11.35}$$

and the corrector step reads

$$f_i^{n+1} = \frac{1}{2}\left[f_i^n + f_i^* - C\left(f_i^* - f_{i-1}^*\right)\right]. \tag{11.36}$$

The scheme is second-order accurate in time and in space, conditionally stable, with stability condition $C \leqslant 1$. The scheme is commonly used and often is highly recommended. For best results, the order of differentiation can be alternated.

11.4   Nonlinear Problems

The previously discussed methods can be applied to the nonlinear problems too. For instance,

$$\frac{\partial u}{\partial t} + u\frac{\partial u}{\partial x} = 0 \tag{11.37}$$

is a nonlinear advection equation representing one-dimensional inviscid flow, which can be rewritten in a conservative form as

$$\frac{\partial u}{\partial t} = \frac{1}{2}\frac{\partial u^2}{\partial x}. \tag{11.38}$$

11.4.1   Explicit methods

Applying the **Lax explicit** method yields

$$\frac{u_i^{n+1} - 1/2(u_{i+1}^n + u_{i-1}^n)}{\Delta t} = -\frac{E_{i+1}^n - E_{i-1}^n}{2\Delta x}, \tag{11.39}$$

where $E = u^2/2$. That is,

$$u_i^{n+1} = \frac{1}{2}(u_{i+1}^n + u_{i-1}^n) - \frac{\Delta t}{4\Delta x}\left[(u_{i+1}^n)^2 - (u_{i-1}^n)^2\right]. \tag{11.40}$$

The solution is stable for $C \leqslant 1$, where the Courant number is defined now by $C = |u_{\max}\Delta t/\Delta x|$.

Applying the **MacCormack** method to the nonlinear advection equation gives

$$u_i^* = u_i^* - \frac{\Delta t}{\Delta x}\left(E_{i+1}^n - E_i^n\right) \tag{11.41}$$

and

$$u_i^{n+1} = \frac{1}{2}\left[u_i^n + u_i^* - \frac{\Delta t}{\Delta x}\left(E_i^* - E_{i-1}^*\right)\right]. \tag{11.42}$$

Stability requirement is that $C = |u_{\max}\Delta t/\Delta x|$ should be less than or equal to unity.

11.4.2   Implicit first-order upwind scheme

For a positive $u > 0$, the implicit scheme can be written as

$$\frac{u_i^{n+1} - u_i^n}{\Delta t} = -\frac{1}{2\Delta x}\left[(u_i^{n+1})^2 - (u_{i-1}^{n+1})^2\right]. \qquad (11.43)$$

It is possible to linearize the above equation:

$$\frac{u_i^{n+1} - u_i^n}{\Delta t} = -\frac{1}{2\Delta x}\left[u_i^n u_i^{n+1} - u_{i-1}^n u_{i-1}^{n+1}\right]. \qquad (11.44)$$

11.5   Runge–Kutta Method

Fourth-order Runge–Kutta (RK) method is as follows:

$$u_i^1 = u_i^n, \qquad (11.45)$$

$$u_i^2 = u_i^n - \frac{\Delta t}{2}\left(\frac{\partial E}{\partial x}\right)_i^1, \qquad (11.46)$$

$$u_i^3 = u_i^n - \frac{\Delta t}{2}\left(\frac{\partial E}{\partial x}\right)_i^2, \qquad (11.47)$$

$$u_i^4 = u_i^n - \frac{\Delta t}{2}\left(\frac{\partial E}{\partial x}\right)_i^3. \qquad (11.48)$$

Finally,

$$u_i^{n+1} = u_i^n - \frac{\Delta t}{6}\left[\left(\frac{\partial E}{\partial x}\right)_i^1 + 2\left(\frac{\partial E}{\partial x}\right)_i^2 + 2\left(\frac{\partial E}{\partial x}\right)_i^3 + \left(\frac{\partial E}{\partial x}\right)_i^4\right].$$
$$(11.49)$$

The scheme is explicit, fourth-order accurate in time and is conditionally stable ($C \leqslant 2\sqrt{2}$). In order to save computer memory, the above equations can be written as

$$u_i^1 = u_i^n, \qquad (11.50)$$

$$u_i^2 = u_i^n - \frac{\Delta t}{4}\left(\frac{\partial E}{\partial x}\right)_i^1, \qquad (11.51)$$

$$u_i^3 = u_i^n - \frac{\Delta t}{3} \left(\frac{\partial E}{\partial x} \right)_i^2, \qquad (11.52)$$

$$u_i^4 = u_i^n - \frac{\Delta t}{2} \left(\frac{\partial E}{\partial x} \right)_i^3, \qquad (11.53)$$

$$u_i^{n+1} = u_i^n - \Delta t \left(\frac{\partial E}{\partial x} \right)_i^4. \qquad (11.54)$$

Note that using central difference to approximate in space leads to spurious oscillation.

11.6   Advection–Diffusion Problems

Unsteady, one-dimensional parabolic advection–diffusion equation can be written as

$$\frac{\partial f}{\partial t} + u \frac{\partial f}{\partial x} = \lambda \frac{\partial^2 f}{\partial x^2}. \qquad (11.55)$$

11.6.1   Central difference in space scheme

The above equation can be approximated using forward difference in time and central difference in space. The explicit scheme is

$$f_i^{n+1} = f_i^n - \frac{C}{2}(f_{i+1}^n - f_{i-1}^n) + D(f_{i+1}^n - 2f_i^n + f_{i-1}^n), \qquad (11.56)$$

where $C = u\Delta t/\Delta x$ and $D = \lambda \Delta t/\Delta x^2$.
The stability analysis of the scheme yields

$$G = (1 - 2D) + 2D \cos \theta - IC \sin \theta. \qquad (11.57)$$

Hence, the scheme is stable for $C^2 \leqslant 2D \leqslant 1$.
The implicit scheme reads

$$-\left(\frac{C}{2} + D \right) f_{i-1}^{n+1} + (1 + 2D)f_i^{n+1} + \left(\frac{C}{2} - D \right) f_{i+1}^{n+1} = f_i^n,$$

$$(11.58)$$

for which the stability analysis yields

$$|G|^2 = \frac{1}{|1 + 2D(1 - \cos\theta) + IC\sin\theta|^2}$$

$$= \frac{1}{1 + \left[4D^2(1 - \cos\theta)^2 + C^2\sin^2\theta + 4D(1 - \cos\theta)\right]}. \quad (11.59)$$

Obviously, $|G|^2 \leqslant 1$ for all $\theta \in (0;\ 2\pi)$ regardless of the values of C and D. So, the scheme is unconditionally stable.

11.6.2 MacCormack scheme

The explicit MacCormack scheme can be written as follows: predictor step:

$$f_i^* = f_i^n - \frac{u\Delta t}{\Delta x}\left(f_{i+1}^n - f_i^n\right) + \frac{\lambda\Delta t}{\Delta x^2}\left(f_{i+1}^n - 2f_i^n + f_{i-1}^n\right), \quad (11.60)$$

corrector step:

$$f_i^{n+1} = \frac{1}{2}\left[f_i^n + f_i^* - C\left(f_i^* - f_{i-1}^*\right) + D\left(f_{i+1}^* - 2f_i^* + f_{i-1}^*\right)\right]. \quad (11.61)$$

The stability of this scheme requires that the condition $(C+2D) \leqslant 1$ has to be satisfied. Derivation of the stability condition is left as an exercise.

11.7 Finite Volume Method Applied to Advection Equation

This section deals mainly with nonlinear problems encountered in compressible fluid dynamics when using finite volume method (FVM). Although the extension to any nonlinear system of equations is straightforward, the readers whose major focus is not fluid mechanics problems and related physics can ignore this part safely.

The most general conservative shape of the one-dimensional nonlinear advection equations reads

$$\frac{\partial f}{\partial t} + \frac{\partial F}{\partial x} = 0, \quad (11.62)$$

where f can be a scalar or vector quantity and $F = F(f)$ is the flux (density) function of f. Of course, if f is a scalar, $F(f)$ is also a scalar, and if f is a vector, $F(f)$ is also a vector of the same size. The non-conservative shape can be deduced by applying the chain rule, $\dfrac{\partial F}{\partial x} = A(f)\dfrac{\partial f}{\partial x}$, where $A(f) = \dfrac{\partial F}{\partial f}$ is the Jacobian matrix of F of size $m \times m$ and m is the size of the system (11.62) or, equivalently, the dimension of f. Hence, the non-conservative equation reads as follows:

$$\frac{\partial f}{\partial t} + A\frac{\partial f}{\partial x} = 0. \tag{11.63}$$

For instance, if $f = (f_1, f_2)$ is a two-dimensional vector, then $F = (F_1, F_2)^t$ and A is the matrix given by

$$A = \begin{pmatrix} \dfrac{\partial F_1}{\partial f_1} & \dfrac{\partial F_1}{\partial f_2} \\[2mm] \dfrac{\partial F_2}{\partial f_1} & \dfrac{\partial F_2}{\partial f_2} \end{pmatrix}.$$

For compressible, inviscid flow, Eq. (11.62) is just Euler's equations where $f = (f_1, f_2, f_3)^t = (\rho, \rho u, \rho E)^t$. Here ρ, u, and E are the fluid density, velocity, and specific total energy (i.e., the sum of the internal and kinetic energies per unit of mass), respectively. $F = (\rho u, \rho u^2 + p, \rho u(E + p/\rho))^t$, where p is the pressure. f_1, f_2, and f_3 are called conservative variables, and ρ, u, and p primitive variables.

For hyperbolic problems, such as Euler's equations, matrix A is diagonalizable, i.e., it has exactly n real (not necessarily different) eigenvalues. If Λ is the diagonal matrix with elements as these eigenvalues and R is the matrix with columns as the associated eigenvectors, then it can be shown that

$$A = R\Lambda R^{-1} \tag{11.64}$$

and so that

$$\Lambda = R^{-1}AR.$$

So, multiplication of Eq. (11.63) by R^{-1}, using the identity $A = A R R^{-1}$ (since R necessarily exists for hyperbolic equations and is invertible), and letting $W = R^{-1}f$ with some algebra gives the following equation:

$$\frac{\partial W}{\partial t} + \Lambda \frac{\partial W}{\partial x} = \left(\frac{\partial R^{-1}}{\partial t} + \Lambda \frac{\partial R^{-1}}{\partial x} \right) R\, W. \tag{11.65}$$

If the problem is linear, then this system reduces to

$$\frac{\partial W}{\partial t} + \Lambda \frac{\partial W}{\partial x} = 0, \tag{11.66}$$

which is merely a set of uncoupled differential equations. Each of these equations, being a scalar equation, can therefore be solved separately. Equations (11.65) are called characteristic equations, and W is called the vector of characteristic variables.

11.7.1   Centered schemes

11.7.1.1   *Basic formulation*

The most straightforward scheme is the forward-time centered-space (FTCS) that approximates the numerical flux by

$$F_{i+\frac{1}{2}} = \frac{1}{2} \left(F_i + F_{i+1} \right),$$

where $F_i = F(f_i)$ and $F_{i+1} = F(f_{i+1})$. As mentioned before, using such a centered scheme leads to a second-order truncation error but is unstable unless some artificial dissipation is introduced.

To stabilize the scheme, an artificial explicit dissipation term $D(f)$ is considered in the numerical flux:

$$F_{i+\frac{1}{2}} = \frac{1}{2} \left(F_i + F_{i+1} \right) - D_{i+\frac{1}{2}} \left(f_{i+1} - f_i \right).$$

The coefficient $D(f)$ must satisfy the following two criteria:

* $\forall f : D(f) > 0$;
* $D(f)$ must preserve the second-order accuracy.

11.7.1.2 Jameson–Schmidt–Turkel scheme

Jameson–Schmidt–Turkel scheme modifies Eq. (11.62) as follows:

$$\frac{\partial f}{\partial t} + \frac{\partial F}{\partial x} = \mu \frac{\partial^2 f}{\partial x^2} - \lambda \frac{\partial^4 f}{\partial x^4}, \tag{11.67}$$

where $\mu > 0$ and $\lambda > 0$ are constants. This equation is equivalent to considering a modified flux that would be given by

$$F - \mu \frac{\partial f}{\partial x} + \lambda \frac{\partial^3 f}{\partial x^3}. \tag{11.68}$$

If $\rho(A)$ is the spectral radius of the Jacobian matrix A, then the new flux can be approximated by

$$F_{i+\frac{1}{2}} = \frac{1}{2}\left(F_i + F_{i+1}\right) - \epsilon^{(2)}_{i+\frac{1}{2}} \rho\left(A_{i+\frac{1}{2}}\right)\left(f_{i+1} - f_i\right)$$

$$+ \epsilon^{(4)}_{i+\frac{1}{2}} \rho\left(A_{i+\frac{1}{2}}\right)\left(f_{i+2} - 3f_{i+1} - 3f_i - f_{i-1}\right),$$

where $\rho\left(A_{i+\frac{1}{2}}\right) = \max\left(\rho\left(A_i\right), \rho\left(A_{i+1}\right)\right)$, which corresponds to the maximum wave speed between cells i and $i+1$.

The coefficient $\epsilon^{(2)}$ is effective in the neighborhood of steep variations (shock waves, for example) and is given by

$$\epsilon^{(2)}_{i+\frac{1}{2}} = \kappa_2 \max\left(s_i, s_{i+1}\right),$$

where

$$s_i = \frac{|p_{i+1} - 2p_i + p_{i-1}|}{p_{i+1} + 2p_i + p_{i-1}},$$

where $p > 0$ is the pressure. In regions of smooth pressure variations, s_i and s_{i+1} are small and so is $\epsilon^{(2)}_{i+\frac{1}{2}}$. The coefficient κ_2, $\frac{1}{4} \leqslant \kappa_2 \leqslant \frac{1}{2}$, is set by the user.

The coefficient $\epsilon^{(4)}$ is effective far away from regions of steep variations and is introduced to damp out spurious oscillations of high frequency and so helps reaching steady state quickly. It is given by

$$\epsilon^{(4)}_{i+\frac{1}{2}} = \max\left(0, \kappa_4 - \epsilon^{(2)}_{i+\frac{1}{2}}\right),$$

where the coefficient κ_4, $\frac{1}{64} \leqslant \kappa_2 \leqslant \frac{1}{32}$, is set by the user too. The larger the κ_4, the more diffusion the solution suffers. Hence, the coefficient must be kept as small as possible.

11.7.2   Flux vector splitting schemes

11.7.2.1   *Introduction with a scalar equation*

To illustrate the idea, let us consider again the scalar diffusion equation:

$$\frac{\partial f}{\partial t} + u\frac{\partial f}{\partial x} = 0.$$

If $u > 0$, applying the upwind scheme yields the following approximation of the flux:

$$F_{i+\frac{1}{2}} = (uf)_i.$$

Otherwise,

$$F_{i+\frac{1}{2}} = (uf)_{i+1}.$$

If we ignore the sign of u, we can apply the following approximation regardless of the sign of u:

$$F_{i+\frac{1}{2}} = \left(\frac{u + |u|}{2}f\right)_i + \left(\frac{u - |u|}{2}f\right)_{i+1}.$$

By setting $u^+ = \frac{u+|u|}{2}$ and $u^- = \frac{u-|u|}{2}$, the flux approximation reads

$$F_{i+\frac{1}{2}} = F_{i+\frac{1}{2}}^+ + F_{i+\frac{1}{2}}^- = \frac{1}{2}\left(F_i + F_{i+1} - |u|\left(f_{i+1} - f_i\right)\right), \qquad (11.69)$$

where

$$\begin{aligned} F_{i+\frac{1}{2}}^+ &= \left(u^+ f\right)_i, \\ F_{i+\frac{1}{2}}^- &= \left(u^- f\right)_{i+1}. \end{aligned} \qquad (11.70)$$

11.7.2.2   *Extension to a hyperbolic system of equations*

Generalization of the previous systematic upwind scheme to a hyperbolic system is straightforward, considering the characteristic shape (11.66).

If we denote by $|\Lambda|^{\#}$ the matrix whose elements are the absolute values of the respective elements of matrix Λ, then we can define the following matrices Λ^{+} and Λ^{-}:

$$\Lambda^{+} = \frac{1}{2}\left(\Lambda + |\Lambda|^{\#}\right), \quad \Lambda^{-} = \frac{1}{2}\left(\Lambda - |\Lambda|^{\#}\right).$$

If we put $A^{+} = R\Lambda^{+}R^{-1}$ and $A^{-} = R\Lambda^{-}R^{-1}$, we can easily check that $A = A^{+} + A^{-}$, and in a similar manner to (11.70), we can write

$$F^{+}_{i+\frac{1}{2}} = \left(A^{+}f\right)_{i},$$
$$F^{-}_{i+\frac{1}{2}} = \left(A^{-}f\right)_{i+1}. \tag{11.71}$$

This scheme was first derived by Prendergast but was advocated by Steger and Warming and has been called after their names.

11.7.2.3   *Modified Steger and Warming scheme*

Because of the large error observed when applying their original scheme to viscous fluid flows, a modified Steger and Warming scheme was suggested. The new expression of the fluxes $F^{+}_{i+\frac{1}{2}}$ and $F^{-}_{i+\frac{1}{2}}$ are

$$F^{+}_{i+\frac{1}{2}} = A^{+}_{i+\frac{1}{2}}f^{-}_{i+\frac{1}{2}},$$
$$F^{-}_{i+\frac{1}{2}} = A^{-}_{i+\frac{1}{2}}f^{+}_{i+\frac{1}{2}}, \tag{11.72}$$

where $A^{+}_{i+\frac{1}{2}} = \frac{1}{2}\left(A^{+}_{i} + A^{+}_{i+1}\right)$, $A^{-}_{i+\frac{1}{2}} = \frac{1}{2}\left(A^{-}_{i} + A^{-}_{i+1}\right)$, and $f^{+}_{i+\frac{1}{2}} = f_{i+1}$ and $f^{-}_{i+\frac{1}{2}} = f_{i}$.

11.7.2.4 *Vijayasundaram scheme*

In this scheme,

$$
\begin{aligned}
F^+_{i+\frac{1}{2}} &= A^+(f_{i+\frac{1}{2}})f_i, \\
F^-_{i+\frac{1}{2}} &= A^-(f_{i+\frac{1}{2}})f_{i+1},
\end{aligned}
\tag{11.73}
$$

where $f_{i+\frac{1}{2}} = \frac{1}{2}\left(f_i + f_{i+1}\right)$.

11.7.2.5 *van Leer scheme*

Yet another enhancement of Steger–Warming scheme was proposed by van Leer based on the value of Mach number $M = \frac{u}{c}$, where u is the flow speed and c is the speed of sound. The approximation of the flux according to van Leer has to satisfy at least the following conditions:

- Consistency: $F(f) = F^+ + F^-$;
- The eigenvalues of the splitting Jacobian matrix $\frac{\partial F^+}{\partial f}$ (resp. $\frac{\partial F^-}{\partial f}$) must be positive (resp. negative) real numbers;
- Full upwinding must apply for supersonic flow: for $M > 1$, $F^- = 0$ and for $M < -1$, $F^+ = 0$;
- The Jacobian matrices $\frac{\partial F^+}{\partial f}$ and $\frac{\partial F^-}{\partial f}$ must be continuous;
- F^+ and F^- must preserve symmetry (and antisymmetry) of the flux components, i.e.,

$$
F^+_k(M) = \pm F^-_k(-M) \quad \text{when } F_k(M) = \pm F_k(-M), \quad k = 1, \dots, m;
$$

- $F^+ = F^+(M)$ and $F^- = F^-(M)$ must be polynomials of the lowest possible degree.

A particular scheme was proposed by van Leer himself and is given in Table 11.1. It can be shown that this scheme is first-order accurate and the stability criterion reads

$$
\begin{cases}
\dfrac{(|u| + c)\Delta t}{\Delta x} \leqslant 1 & \text{when } |M| \geqslant 1, \\[2ex]
\dfrac{(|u| + c)\Delta t}{\Delta x} \leqslant \dfrac{\gamma + |M|(3 - \gamma)}{\gamma + 3} & \text{when } |M| \leqslant 1.
\end{cases}
\tag{11.74}
$$

Table 11.1:   van Leer approximation of the fluxes. ρ is the fluid density and γ is the heat capacity ratio.

M	$F^+_{i+\frac{1}{2}}$	$F^-_{i+\frac{1}{2}}$
$M \leqslant -1$	0	F_{i+1}
$-1 \leqslant M \leqslant 1$	$\frac{\rho c}{4}(M+1)^2 \begin{pmatrix} 1 \\ \frac{2c}{\gamma}\left(1+\frac{\gamma-1}{2}M\right) \\ \frac{2c^2}{\gamma^2-1}\left(1+\frac{\gamma-1}{2}M\right)^2 \end{pmatrix}$	$-\frac{\rho c}{4}(M-1)^2 \begin{pmatrix} 1 \\ -\frac{2c}{\gamma}\left(1-\frac{\gamma-1}{2}M\right) \\ \frac{2c^2}{\gamma^2-1}\left(1-\frac{\gamma-1}{2}M\right)^2 \end{pmatrix}$
$M \geqslant 1$	F_i	0

11.7.3   Riemann solvers: Flux difference splitting

11.7.3.1   *Godunov's scheme*

Godunov's scheme relies on solving Riemann's problem. The principle of the approach is as in the following steps:

- Approximate the solution f by a piecewise constant function over each cell;
- Solve at each face $x_{i+\frac{1}{2}}$ separating two contiguous cells x_i and x_{i+1} the Riemann's problem corresponding to the jump (discontinuity) introduced at that face. The flux $F_{i+\frac{1}{2}}$ is computed. This is done for all faces;
- Update the solution using that flux and setting

$$f_i^{n+1} = f_i^n - \frac{\Delta t}{\Delta x}\left(F_{i+\frac{1}{2}} - F_{i-\frac{1}{2}}\right).$$

To illustrate the method, let us consider the following Riemann's problem:

$$\frac{\partial f}{\partial t} + A\frac{\partial f}{\partial x} = 0,$$

with the following initial conditions:

$$\begin{cases} f_L & \text{if } x < 0, \\ f_R & \text{if } x > 0. \end{cases} \tag{11.75}$$

Since the problem is hyperbolic, matrix A is diagonalizable with real eigenvalues λ_k, $k = 1, \ldots, m$ (m the size of f). As before, let R be the matrix of the eigenvectors of A and R_k the (columns) eigenvector associated with eigenvalue λ_k, and let the vector $W = R^{-1}f$ of elements w_k, $k = 1, \ldots, m$. It is easy to check that the solution to this problem reads

$$f(x, t) = \sum_{k=1}^{m} f_k(x - \lambda_k t, 0) R_k.$$

Using the basis $(R_k)_{k=1,\ldots,m}$, we can always write

$$f_L = \sum_{k=1}^{m} \alpha_k R_k \quad \text{and} \quad f_R = \sum_{k=1}^{m} \beta_k R_k,$$

where $\alpha_k = f_L \cdot R_k$ and $\beta_k = f_R \cdot R_k$. The solution can be recast therefore as follows:

$$f(x, t) = f_L + \sum_{\lambda_k < x/t} (\beta_k - \alpha_k) R_k = f_R - \sum_{\lambda_k > x/t} (\beta_k - \alpha_k) R_k.$$

Hence, the flux at face $x = 0$ (we label $x_{i+\frac{1}{2}}$) reads

$$F_{i+\frac{1}{2}} = \frac{1}{2} \left(F_R + F_K - |A|(f_R - f_L) \right). \tag{11.76}$$

Recall that $|A|$ is the determinant of matrix A.

It can shown that this scheme is first-order accurate and the stability criterion is $\frac{(|u|+c)\Delta t}{\Delta x} \leqslant 1$.

11.7.3.2   Roe's scheme

In Roe's scheme, the nonlinear problem is linearized as follows:

$$\frac{\partial f}{\partial t} + \tilde{A}\frac{\partial f}{\partial x} = 0,$$

where $\tilde{A}$ depends only on f_L and f_R and is assumed constant in the time interval $(n\Delta t; (n+1)\Delta t)$.

Roe defines the matrix in a way to satisfy the constraints of:

- conservativeness: $\tilde{A} \times (f_R - f_L) = F(f_R) - F(f_L)$;
- diagonalizable with real eigenvalues;
- consistency.

He showed that the matrix $\tilde{A}$ is identical to the matrix $A(\tilde{f})$ calculated for an average state $\tilde{f}$ called Roe-average, defined by

$$a = \sqrt{\frac{\rho_R}{\rho_L}}, \qquad \tilde{\rho} = a\rho_L,$$

$$\tilde{u} = \frac{u_L + au_R}{1 + a}, \qquad \tilde{H} = \frac{E_L + aH_R}{1 + a},$$

where $H = E + \frac{p}{\rho}$. The approximation of the flux reads under these conditions

$$F_{i+\frac{1}{2}} = \frac{1}{2}\left(F_i + F_{i+1} - |\tilde{A}_{i+\frac{1}{2}}|(f_{i+1} - f_i)\right). \tag{11.77}$$

This scheme is first-order accurate and is stable under the criterion $\frac{(|u|+c)\Delta t}{\Delta x} \leqslant 1$. It resolves shocks accurately. However, expansion shocks may happen which makes Roe's scheme non-entropic. An entropy correction can be added to correct this misbehavior.

Harten proposes to replace the absolute value of the eigenvalues when they approach zero with a nonzero quadratic function. Thus, for each eigenvalue λ_i, the quantity defined by

$$|\lambda_i|_\delta = \begin{cases} |\lambda_i| & \text{if } |\lambda_i| < \delta, \\ \frac{1}{2}\left(\frac{\lambda_i^2}{\delta} + \delta\right) & \text{if } |\lambda_i| \geqslant \delta \end{cases} \tag{11.78}$$

is used. The parameter δ must be tuned according to the flow conditions. This additional dissipation affects the accuracy of the results, so it is desirable to reduce the magnitude of this parameter as much as possible. With Harten's entropy correction, the approximation of the flux flow reads

$$F_{i+\frac{1}{2}} = \frac{1}{2}\left(F_i + F_{i+1} - \tilde{R}|\Lambda_i|_\delta \tilde{R}^{-1}(f_{i+1} - f_i)\right). \tag{11.79}$$

11.7.3.3   *Harten, Lax, and van Leer schemes*

Some particular schemes based on solving an approximate Riemann problem were derived by Harten, Lax, and van Leer (HLL). The originality is that the computation of intermediate states is performed over integrals across the different waves considered.

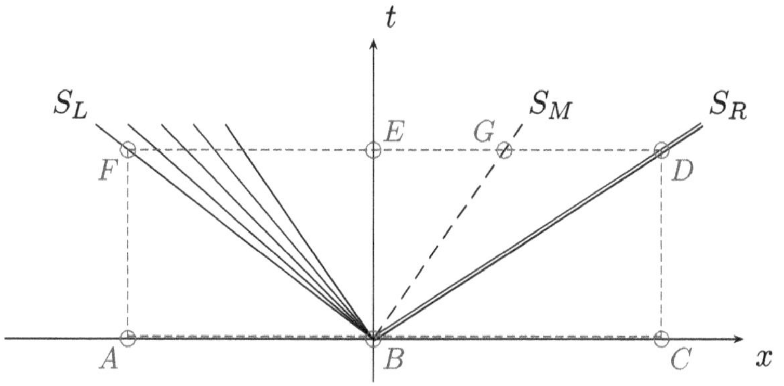

Figure 11.1: Exact Riemann's problem. The volume taken into consideration is represented in dashed line.

11.7.3.3.1 Integral formulation of Riemann's problem: Let us consider the complete Riemann's problem depicted in Fig. 11.1. Let S_L and S_R be the maximum (in absolute value) speeds of waves generated at the face to the left and to the right of the latter, respectively. In Figure 11.1, the case considered is such that $S_L < 0 < S_R$. These speeds are the (signed) slope of the wave characteristic curve. If we denote $\tilde{f}_L$, $\tilde{f}_R$, and f^* as the mean values of the state on segments $[EF]$, $[ED]$, and $[FD]$, respectively, then we can write

$$(S_R - S_L)f^* = S_R \tilde{f}_R - S_L \tilde{f}_L.$$

We denote by F^* the mean flux at $x = 0$ along the segment $[EB]$. In Godunov's scheme, it is merely the approximation of the numerical flux.

Integrating the advection equations over domains $[ACDF]$, $[BCDE]$, and $[ABEF]$ leads to the following relationships:

$$F(f_R) - F(f_L) = S_R f_R - S_L f_L - (S_R - S_L)f^*, \quad (11.80)$$

$$F(f_R) - F^* = S_R(f_R - \tilde{f}_R), \quad (11.81)$$

$$F(f_L) - F^* = S_L(f_L - \tilde{f}_L). \quad (11.82)$$

The combination of these relationships yields the following expression of the flux F^*:

$$F^* = \frac{1}{2}\left(F(f_L) + F(f_R) - S_R(f_R - \tilde{f}_R) - S_L(f_L - \tilde{f}_L)\right). \quad (11.83)$$

The point is that more unknowns are involved than equations: four unknowns, f_L, f_R, $\tilde{f}_L$, and $\tilde{f}_R$ against three Eqs. (11.80), (11.81), and (11.82). The last equation to be provided is a relationship between $\tilde{f}_L$ and $\tilde{f}_R$.

11.7.3.3.2   Harten, Lax, and van Leer basic scheme: HLL considers Riemann's problem with exactly two waves separating a domain with uniform state f^*, see Fig. 11.2. The additional relationship to be considered is merely the equality of the states $\tilde{f}_L$ and $\tilde{f}_R$. So, Eq. (11.80) yields the following expression of f^*:

$$f^* = \frac{S_R f_R - S_L f_L - (F(f_R) - F(f_L))}{S_R - S_L}. \qquad (11.84)$$

Introducing the latter expression into (11.83) gives

$$F^* = \frac{S_R F_L - S_L f_R + S_R S_L \, (f_R - f_L)}{S_R - S_L}.$$

What remains is to approximate the speeds S_L and S_R. Many options are available, as follows:

- Isentropic wavefront:

$$S_L = \min(0, f_L - c_L) \quad \text{and} \quad S_R = \max(0, f_R + c_R);$$

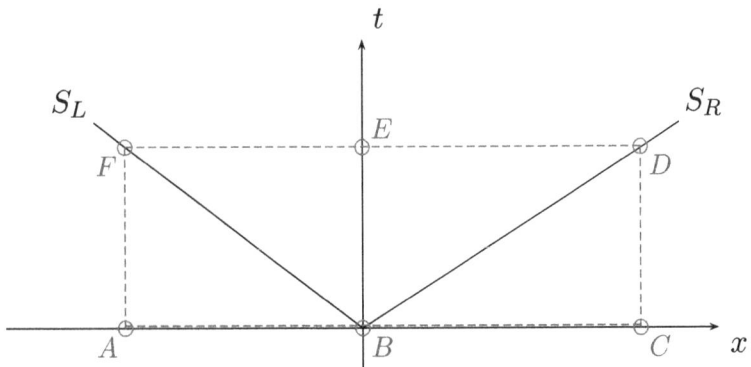

Figure 11.2:   Two-wave Riemann's problem. The volume taken into consideration is represented in dashed line.

- Mean Roe's state, called HLLR scheme:

$$S_L = \min(0, \tilde{f} - \tilde{c}) \quad \text{and} \quad S_R = \max(0, \tilde{f} + \tilde{c}),$$

where $\tilde{f}$ and $\tilde{c}$ are the velocity and speed of sound, respectively, of Roe's mean state;
- Einfeldt scheme, called HLLE:

$$S_L = \min(0, \tilde{f} - \tilde{c}, f_L - c_L) \quad \text{and} \quad S_R = \max(0, \tilde{f} + \tilde{c}, f_R + c_R).$$

The ultimate expression of the flux reads

$$F_{i+\frac{1}{2}} = \begin{cases} F(f_L) & \text{if } S_L > 0, \\ \dfrac{S_R F_L - S_L f_R + S_R S_L \left(f_R - f_L\right)}{S_R - S_L} & \text{if } S_L \leqslant 0 \leqslant S_R, \\ F(f_R) & \text{if } S_R < 0. \end{cases}$$

$$\tag{11.85}$$

This scheme is first-order accurate, entropic, and stable for the criterion $\frac{(|u|+c)\Delta t}{\Delta x} \leqslant 1$.

11.7.3.3.3   Harten, Lax, and van Leer contact scheme: In order to represent Riemann's problem even closer, three waves are taken into account. Two different states separated by the contact discontinuity (density discontinuity) f_L^* and f_R^* are considered. These states are equal to the mean value of the solution of Riemann's problem on segments $[FG]$ and $[GD]$, respectively, see Fig. 11.1.

Assuming f to be constant in each region, the expression of the flux reads

$$F_{i+\frac{1}{2}} = \frac{1}{2}\left(F(f_L) + F(f_R)\right) - \frac{1}{2}\left(S_R(f_R - f_R^*)\right)$$
$$+ |S_M|(f_R^* - f_L^*) + S_L(f_L^* - f_L)). \tag{11.86}$$

Integrating the advection equations over domains $[ACDF]$, $[BCDE]$, and $[ABEF]$ leads to the following relationships:

$$F(f_L^*) - F(f_L) = S_L(f_L^* - f_L), \tag{11.87}$$
$$F(f_R^*) - F(f_R) = S_R(f_R^* - f_R), \tag{11.88}$$
$$F(f_R^*) - F(f_L^*) = S_M(f_R^* - f_L^*). \tag{11.89}$$

It is possible from these three relations to compute intermediate states. In order to satisfy relation (11.89) systematically and since the wave whose speed is S_M is a contact discontinuity, i.e., a wave across which only the density experiences a jump, we set

$$u_R^* = u_L^* = S_M \quad \text{and} \quad P_R^* = P_L^* = P^*.$$

This expresses only the continuity of the velocity and pressure across this wave. The densities ρ_L^* and ρ_R^* are computed using Rankine–Hugoniot jump conditions. Once more, we have more unknowns than equations. According to Toro and Batten, if we assume the continuity of the energy, the continuity of the pressure yields the following expression of S_M:

$$S_M = \frac{P_R - P_L + \rho_L u_L (S_L - u_L) - \rho_R u_R (S_R - u_R)}{\rho_L (S_L - u_L) - \rho_R (S_R - u_R)}$$

and the expression of the pressure P^*:

$$P_{i+\frac{1}{2}} = P^* = P_L + \rho_L (u_L - S_L)(u_L - S_M) = P_R + \rho_R (u_R - S_R)(u_R - S_M).$$

We get then all the quantities f_K^* ($K = L,\ R$):

$$f_K^* = \begin{pmatrix} \rho_K^* \\ (\rho u)_K^* \\ (\rho E)_K^* \end{pmatrix} = \frac{1}{S_K - S_M} \begin{pmatrix} \rho_K^* (S_K - S_M) \\ (\rho u)_K (S_K - u_K) + P^* - P_K \\ (\rho E)_K (S_K - u_K) + P^* S_M P_K u_K \end{pmatrix}. \tag{11.90}$$

The flux F at intermediate states f_K^* reads

$$F(f_K^*) = \begin{pmatrix} \rho_K^* S_M \\ (\rho u)_K^* S_M + P^* \\ (\rho E)_K^* S_M + P^* S_M \end{pmatrix}. \tag{11.91}$$

The approximation of the flux at the face therefore reads

$$F_{i+\frac{1}{2}} = \begin{cases} F(f_L) & \text{if } S_L > 0, \\ F(f_L^*) & \text{if } S_L \leqslant 0 \leqslant S_M, \\ F(f_R^*) & \text{if } S_M \leqslant 0 \leqslant S_R, \\ F(f_L) & \text{if } S_R < 0. \end{cases} \tag{11.92}$$

In the same fashion, S_L and S_R have to be approximated. The options provided in Section 11.7.3.3.2 can be used, but Toro suggested another option for his own Harten, Lax, and van Leer contact (HLLC) scheme ($K = L$, R):

$$S_K = u_K - c_L q_K,$$

where q_K is given by

$$q_K = \begin{cases} 1 & \text{if } P^* \leqslant P_K, \\[3ex] \dfrac{1}{\sqrt{1 + \dfrac{\gamma + 1}{2\gamma}\left(\dfrac{P^*}{P_K} - 1\right)}} & \text{if } P^* > P_K. \end{cases}$$

11.7.4 Liou and Steffen AUSM scheme and its variant

The flux vector splitting (FVS) schemes are particularly robust and are able to capture strong flow discontinuities like shocks and expansion waves but are too diffusive for viscous fluid flows. On the other hand, the flux difference splitting (FDS) schemes deal correctly with the boundary layer problems (and so viscous fluid flows) but usually experience stability issues and may even compute non-entropic solutions. The advection upstream splitting methods (AUSMs) are based on the idea of combining FVS and FDS schemes in order to take advantage of both. In other words, AUSMs intend to preserve the robustness of the decomposition schemes and the accuracy of Riemann's solvers at the same time.

11.7.4.1 *Liou's basic AUS method*

The basic principle of the AUSM originally proposed by Liou and Steffen is to consider acoustic waves and fluid convection as two separate mechanisms. The flux of Euler's equations is first written in the following form:

$$F = \begin{pmatrix} \rho u \\ \rho u^2 + p \\ \rho H u \end{pmatrix} = \underbrace{\begin{pmatrix} \rho u \\ \rho u^2 \\ \rho H u \end{pmatrix}}_{\text{convection } F_c} + \underbrace{\begin{pmatrix} 0 \\ p \\ 0 \end{pmatrix}}_{\text{acoustics}}. \tag{11.93}$$

The convective flux F_c is written as follows: $F_c = M\,c\Phi$, where $M = \frac{u}{c}$ is the signed Mach number. So, the approximation of the flux at face $i + \frac{1}{2}$ reads

$$F_{i+\frac{1}{2}} = M_{i+\frac{1}{2}} c_{i+\frac{1}{2}} \Phi_{i+\frac{1}{2}} + p_{i+\frac{1}{2}}.$$

The term $\Phi_{i+\frac{1}{2}}$ is calculated by applying the upwind scheme according to the Mach number sign. Like the scheme introduced in Section 11.7.2.1, the Mach number is written as follows:

$$M_{i+\frac{1}{2}} = M_i^+ + M_{i+1}^-.$$

Liou suggested to define $M^\pm$ polynomials that have to:

- be consistent;
- be monotonously increasing functions of M;
- be continuously differentiable;
- preserve symmetry of the flux F, see Section 11.7.2.5.

A particular shape of M^+ and M^- that satisfies these requirements is the following:

$$M^\pm = \begin{cases} \dfrac{1}{2}\left(M \pm |M|\right) & \text{if } |M| > 1, \\ \pm\dfrac{1}{4}(M \pm 1)^2 & \text{otherwise.} \end{cases} \tag{11.94}$$

Once $M_{i+\frac{1}{2}}$ is computed, the speed of sound is computed using a full upwind scheme:

$$c_{i+\frac{1}{2}} = \begin{cases} c_i & \text{if } M_{i+\frac{1}{2}} > 1, \\ c_{i+1} & \text{otherwise.} \end{cases} \tag{11.95}$$

Likewise, the pressure is written by Liou as

$$p_{i+\frac{1}{2}} = P_i^+ p_i + P_{i+1}^- p_{i+1},$$

where $P^\pm$ have to:

- be consistent;
- be monotonously increasing functions of M;
- be continuously differentiable;

- preserve positiveness of the pressure;
- preserve symmetry of the flux F, see Section 11.7.2.5,

which leads to

$$P^{\pm} = \begin{cases} \dfrac{1}{2}\left(1 \pm \dfrac{M}{|M|}\right) & \text{if } |M| \geqslant 1, \\[2mm] \dfrac{1}{4}\left(M \pm 1\right)^2 \left(2 \mp M\right) & \text{otherwise.} \end{cases} \tag{11.96}$$

This scheme is first-order accurate and resolves exactly the problems with stationary slip surfaces.

11.7.4.2 Liou's AUSM+

An updated scheme called AUMS+ was also given by Liou, where the shape of M^+ and M^- that satisfies the same requirements is the following:

$$M^{\pm} = \begin{cases} \dfrac{1}{2}\left(M \pm |M|\right) & \text{if } |M| > 1, \\[2mm] \pm\dfrac{1}{4}(M \pm 1)^2 & \text{otherwise.} \end{cases} \tag{11.97}$$

The speed of sound is computed using the geometric mean:

$$c_{i+\frac{1}{2}} = \sqrt{c_i\, c_{i+1}}. \tag{11.98}$$

Finally, functions $P^{\pm}$ are given by

$$P^{\pm} = \begin{cases} \dfrac{1}{2}\dfrac{M \pm |M|}{M} & \text{if } |M| \geqslant 1, \\[2mm] \pm\dfrac{1}{4}\left(M \pm 1\right)^2 \left(\pm 2 - M + \dfrac{3}{4}M\left(M \mp 1\right)^2\right) & \text{otherwise.} \end{cases} \tag{11.99}$$

This scheme is also first-order accurate.

11.7.5 Higher-order schemes

First-order schemes are simple to implement and they present interesting properties such as monotonicity and total variation diminishing (TVD) which will be discussed further in Section 11.7.5.2.

However, their accuracy is generally inadequate for most practical cases. Hence, it is necessary to derive higher-order schemes (at least of second-order accuracy) to remedy the drawbacks that can take place, mainly close to discontinuities and slip surfaces. However, mere substitution by second-order developments in first-order schemes leads to the appearance of oscillations at the discontinuities, analogous to those encountered in centered scheme. Moreover, it has been shown that any second-order linear upwind scheme produces oscillatory solutions. It can be concluded therefore that the only way to avoid these oscillations in a high-order scheme is to introduce nonlinearities.

The most general formulation of a time-forward finite volume method applied to the advection equation reads

$$f_i^{n+1} = f_i^n + \frac{\Delta t}{\Delta x} \left(F_{i+\frac{1}{2}}^n - F_{i-\frac{1}{2}}^n \right).$$

The flux $F_{i+\frac{1}{2}}$ depends on the value of the state variable f (which can be a scalar or a vector) on the left and right of the interface $i + \frac{1}{2}$ that we denote by $f_{i+\frac{1}{2}}^L$ and $f_{i+\frac{1}{2}}^R$, respectively. One can choose any first-order approximation of $F_{i+\frac{1}{2}}$, such as Lax–Friedrichs, Roe, or Liou, based on $f_{i+\frac{1}{2}}^L$ and $f_{i+\frac{1}{2}}^R$ instead of f_i and f_{i+1}.

So, at first order, $f_{i+\frac{1}{2}}^L$ and $f_{i+\frac{1}{2}}^R$ read

$$f_{i+\frac{1}{2}}^L = f_i \quad \text{and} \quad f_{i+\frac{1}{2}}^R = f_{i+1}.$$

Higher-order evaluation is carried out by taking Taylor series expansion of either the flux F itself or the state variable f that the flux depends on. The latter approach will be discussed.

11.7.5.1   *Monotonic upstream-centered scheme for conservation law*

The so-called monotonic upstream-centered scheme for conservation law (MUSCL) was derived by van Leer. In MUSCL reconstruction, the state variable f, rather than being assumed constant, is assumed piecewise linear in each cell. Thus, the point is to reconstruct values

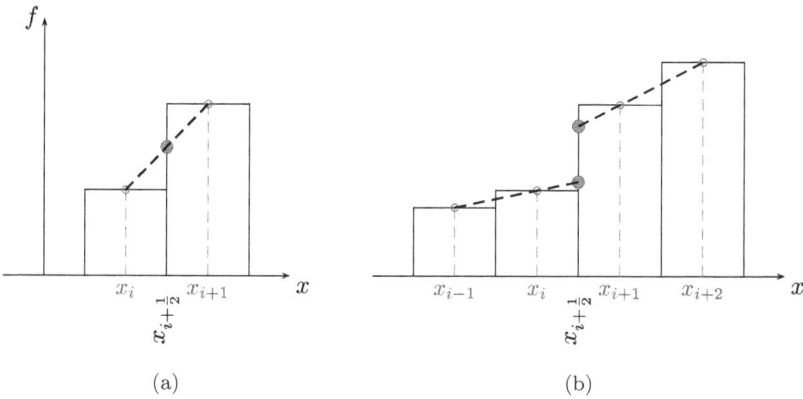

Figure 11.3:   (a) Centered and (b) forward-backward schemes.

at the face $i + \frac{1}{2}$ starting from discrete mean value f_i^n in each cell center. A second-order expansion of the variables is used in order to get the slope. Two options are possible that we can use for uniform mesh size, see Fig. 11.3:

- centered scheme:

$$f_{i+\frac{1}{2}}^L = f_i + \frac{f_{i+1} - f_i}{2},$$

$$f_{i+\frac{1}{2}}^R = f_{i+1} - \frac{f_{i+1} - f_i}{2}.$$

- forward-backward scheme:

$$f_{i+\frac{1}{2}}^L = f_i + \frac{f_i - f_{i-1}}{2},$$

$$f_{i+\frac{1}{2}}^R = f_{i+1} - \frac{f_{i+2} - f_{i+1}}{2}.$$

It is possible to combine both schemes by introducing a skew parameter ϕ:

$$\begin{cases} f_{i+\frac{1}{2}}^L = f_i + \dfrac{1-\phi}{4}\left(f_i - f_{i-1}\right) + \dfrac{1+\phi}{4}\left(f_{i+1} - f_i\right), \\ f_{i+\frac{1}{2}}^R = f_{i+1} - \dfrac{1+\phi}{4}\left(f_{i+1} - f_i\right) - \dfrac{1-\phi}{4}\left(f_{i+2} - f_{i+1}\right). \end{cases} \tag{11.100}$$

It may easily be checked that the scheme is:

- second-order centered if $\phi = 1$;
- second order forward-backward if $\phi = -1$;
- third order if $\phi = \frac{1}{3}$.

The remarkable feature is that this reconstruction can equally be applied to characteristic or primitive variables. For positiveness issues, conservative variables are not used in MUSCL reconstruction but can of course always be calculated from the former sets of variables.

11.7.5.2 *TVD schemes and limiters*

MUSCL reconstruction may induce spurious oscillations close to discontinuities and slip surfaces. Instead of introducing artificial viscosity, another mechanism can be used: the TVD principle. Harten introduced it to avoid the occurrence of non-physical oscillations. *Stricto sensu*, this notion is valid for scalar linear equations. However, in practice, it is applied to systems of linear and nonlinear advection equations such as Euler's equations as well. So, one must always keep in mind that the properties established for scalar cases may not be guaranteed in the case of systems.

For a scalar advection equation valid on a domain D, the total variation of the solution $f(x,t)$ at time τ, denoted $TV(f,\tau)$, is given by

$$TV(f,\tau) = \lim_{\epsilon \to 0} \frac{1}{\epsilon} \int_D |f(x+\epsilon,\tau) - f(x,\tau)| \, dx. \qquad (11.101)$$

When f is differentiable everywhere, that is when no discontinuity happens, then (11.101) becomes

$$TV(f,\tau) = \int_D \left| \frac{\partial f}{\partial x}(x,\tau) \right| dx.$$

A continuous solution w of the advection equation is claimed to be TVD if

$$\forall(t_1, t_2): \quad t_2 \geqslant t_1 \;\Rightarrow\; TV(f,t_2) \leqslant TV(f,t_1).$$

For a discrete solution f_i^n (the approximation of the state $f(x_i, n\Delta t)$), an equivalent definition of total variation, denoted $TV(f^n)$, is given:

$$TV(f^n) = \sum_i \left| f_{i+1}^n - f_i^n \right|.$$

The discrete solution f_i^n of the advection equation is claimed to be TVD if

$$\forall n: \quad TV(f^{n+1}) \leqslant TV(f^n).$$

A numerical scheme is called TVD if the discrete solution it gives preserves the TVD property of the continuous solution.

The main consequence of this feature is that a locally gradient increase is necessarily balanced by a decrease in the gradient elsewhere with at least the same amount. The occurrence of oscillations close to the discontinuity is just not possible with a TVD scheme as long as linear problems are at hand.

MUSCL reconstruction intrinsically introduces nonlinearities in the form of slope limiters. These limiters control the variations in the slopes of the variables on a given cell, used to compute the flux at cell face, by comparing these slopes to those of the neighboring cells. The role of the slope limiter is to decrease locally the scheme accuracy to first order in order to avoid the occurrence of spurious oscillations. So, the truncation error of the overall scheme is never exactly second order. The limiter takes the form of a function $\Psi(r)$, where r denotes the variable slope ratio.

Ultimately, MUSCL reconstruction reads

$$\begin{cases} f_{i+\frac{1}{2}}^L = f_i + \dfrac{1-\phi}{4}\Psi(r^L)\,(f_i - f_{i-1}) + \dfrac{1+\phi}{4}\Psi\left(\dfrac{1}{r^L}\right)(f_{i+1} - f_i), \\[3mm] f_{i+\frac{1}{2}}^R = f_{i+1} - \dfrac{1+\psi}{4}\Psi(r^R)\,(f_{i+1} - f_i) - \dfrac{1-\phi}{4}\Psi\left(\dfrac{1}{r^R}\right)(f_{i+2} - f_{i+1}), \end{cases}$$

$$(11.102)$$

where r^L and r^R are the ratios of the slopes on the left and right of face $i + \frac{1}{2}$, i.e.,

$$r^L = \frac{f_{i+1} - f_i}{f_i - f_{i-1}} \quad \text{and} \quad r^R = \frac{f_{i+1} - f_i}{f_{i+2} - f_{i+1}}.$$

For any scheme to be non-oscillating and at least first order, any limiter $\Psi(r)$ must satisfy at least the following conditions:

- if $0 \leqslant r \leqslant 1$, then $r \leqslant \Psi(r) \leqslant 2r$;
- if $1 \leqslant r \leqslant 2$, then $1 \leqslant \Psi(r) \leqslant r$;
- if $r > 2$, then $1 \leqslant \Psi(r) \leqslant 2$.

Often, the limiters are also asked to satisfy a symmetry constraint which reads

$$\frac{1}{r}\Psi(r) = \Psi\left(\frac{1}{r}\right).$$

Many limiters can be found in the literature, and a few among them are given in Table 11.2. From a practical standpoint, to a given limiter, the numerator $f_{i+1} - f_i$ and the denominator $f_i - f_{i-1}$ (resp. $f_{i+2} - f_{i+1}$) of r^L (resp. r^R) are provided separately instead of providing r^L or r^R themselves. This avoids the possible division-by-zero issue. So, $\Psi = \Psi(a, b)$, where $\frac{a}{b} = r$.

For a nonuniform mesh, the differences involved in the variables a and b are replaced by the slopes $\frac{f_i - f_{i-1}}{x_i - x_{i-1}}$, $\frac{f_{i+1} - f_i}{x_{i+1} - x_i}$, and $\frac{f_{i+2} - f_{i+1}}{x_{i+2} - x_{i+1}}$. The associated limiter has to be bounded by minmod and superbee limiters, see Fig. 11.4. Minmod limiter is the most dissipative and robust of all at the same time. Conversely, superbee is less dissipative but may experience some robustness issues.

Table 11.2:   The most popular limiters used in MUSCL reconstruction.

Limiter name	Expression of $\Psi(r)$	Comment
minmod	$\max(0, \min(1, r))$	
van Leer	$\dfrac{r + |r|}{1 + |r|}$	
van Albada	$\max\left(0, \dfrac{r + r^2}{1 + r^2}\right)$	
Osher	$\max(0, \min(\beta, r))$	$1 \leqslant \beta \leqslant 2$
ospre	$\dfrac{3\left(r^2 + r\right)}{2\left(r^2 + r + 1\right)}$	
monotonized central-diff.	$\max\left(0, \min\left(\dfrac{1+r}{2}, 2r\right)\right)$	
superbee	$\max(0, \min(1, 2r), \min(2, r))$	

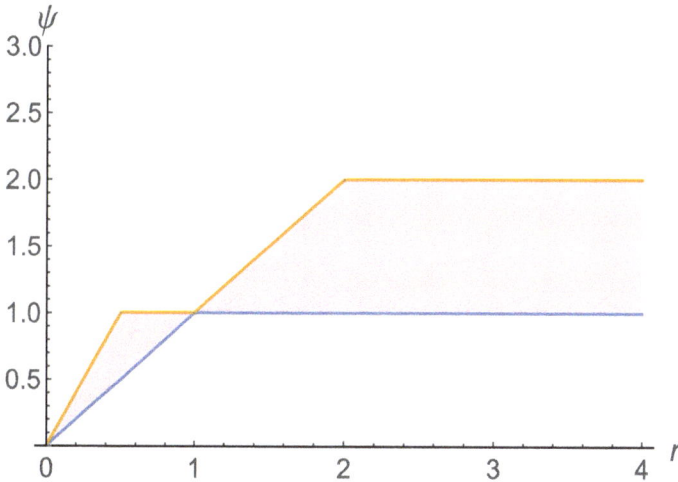

Figure 11.4: Second-order (red-shaded) region encompassed by minmod (blue curve) and superbee (orange curve) limiters. This region is relevant only for MUSCL scheme (11.100).

11.7.6 A worked example

In this example, Sod's shock tube is simulated. A tube is filled with an ideal gas at rest and is divided into two compartments separated by a rigid membrane: a high-pressure compartment with initial conditions $(\rho, u, p) = (1, 0, 1)$ and a low-pressure compartment with initial conditions $(\rho, u, p) = (10^{-1}, 0, 10^{-1})$. The ratio of specific heats of the gas is $\gamma = 1.4$. The tube is 2 units long. At time $t = 0$, the membrane is removed and the gas flows and experiences the propagation of shock and rarefaction (fan) waves.

The tube is divided into 50 cells, and the CFL is set to $\frac{1}{2}$. The high-order MUSCL TVD method, with $\phi = -1$, is used based on the approximation of the flux using Steger and Warming scheme presented in Section 11.7.2.2. Figures 11.5, 11.6, and 11.7 show the effect of different limiters (minmod, van Leer, and superbee, respectively) on the behavior of the solution, especially near discontinuities. The solution presented is the fluid density, ρ, velocity u, pressure p, and dimensionless specific entropy $\frac{\Delta s}{R} = \frac{1}{\gamma - 1} \log \left(\frac{p}{\rho^\gamma} \right)$, where R is the ideal gas constant. As claimed earlier, the minmod limiter is more diffusive, while the superbee limiter leads to more accurate results but experiences some oscillations close to the discontinuities. In this

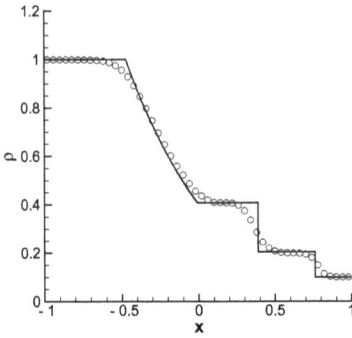

(a) Density

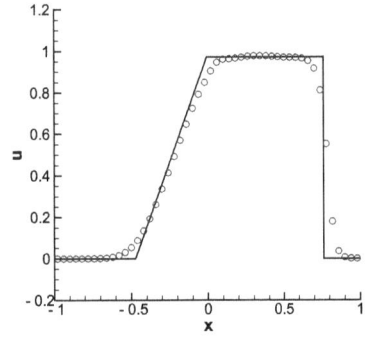

(b) Velocity

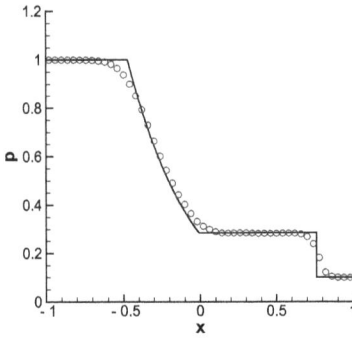

(c) Pressure

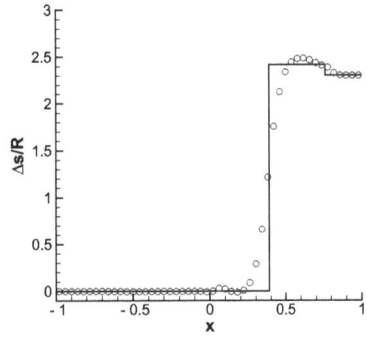

(d) Specific entropy (dimensionless)

Figure 11.5:   Sod's shock tube. Comparison of the exact solution (solid line) and the solution provided using minmod limiter (circles). $t = 0.4$ time units.

case, van Leer limiter shows a good trade-off between non-oscillating and accurate solution.

11.8   Finite Volume Method Applied to Advection–Diffusion Equation

In this section, FVM will be applied to advection–diffusion equation with source term, i.e.,

$$\frac{\partial f}{\partial t} + \frac{\partial (uf)}{\partial x} = \frac{\partial}{\partial x}\left(\lambda \frac{\partial f}{\partial x}\right) + S(t, x). \qquad (11.103)$$

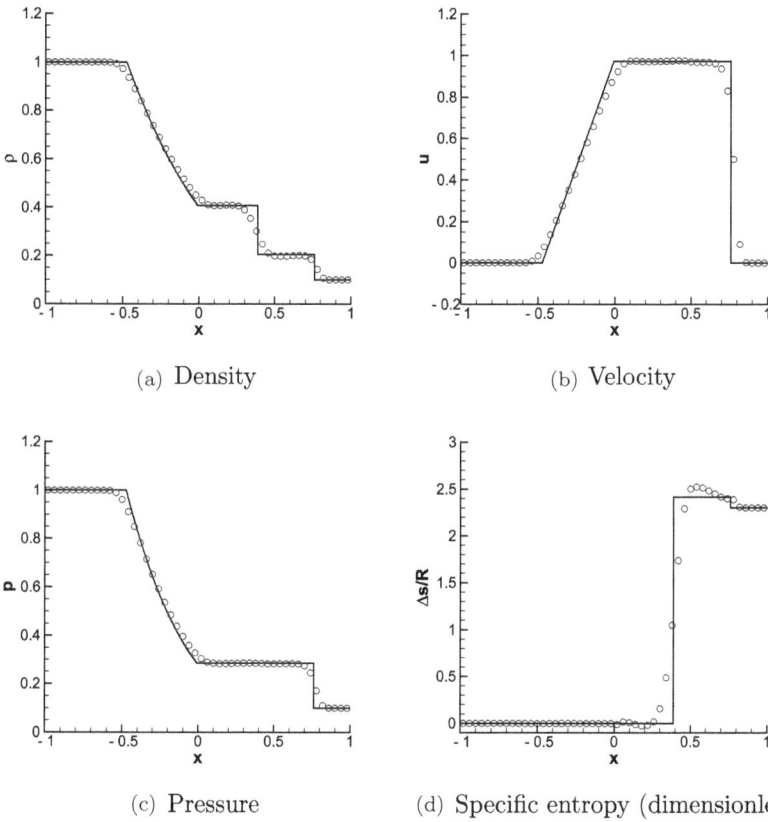

(a) Density

(b) Velocity

(c) Pressure

(d) Specific entropy (dimensionless)

Figure 11.6:   Sod's shock tube. Comparison of the exact solution (solid line) and the solution provided using van Leer limiter (circles). $t = 0.4$ time units.

The above equation is a very general transport equation with variable λ and source term is a function of (t, x). Multiply the above equation by finite domain volume $(\partial t\ \partial x)$ and perform double integration:

$$\int_w^e \int_t^{t+\Delta t} \left(\frac{\partial f}{\partial t} + \frac{\partial (uf)}{\partial x} \right) dx\, dt = \int_t^{t+\Delta t} \int_w^e \frac{\partial}{\partial x} \left(\lambda \frac{\partial f}{\partial x} \right) dx\, dt$$

$$+ \int_t^{t+\Delta t} \int_w^e S(t, x)\, dx\, dt.$$

$$(11.104)$$

Let us deal with the problem term by term.

(a) Density

(b) Velocity

(c) Pressure

(d) Specific entropy (dimensionless)

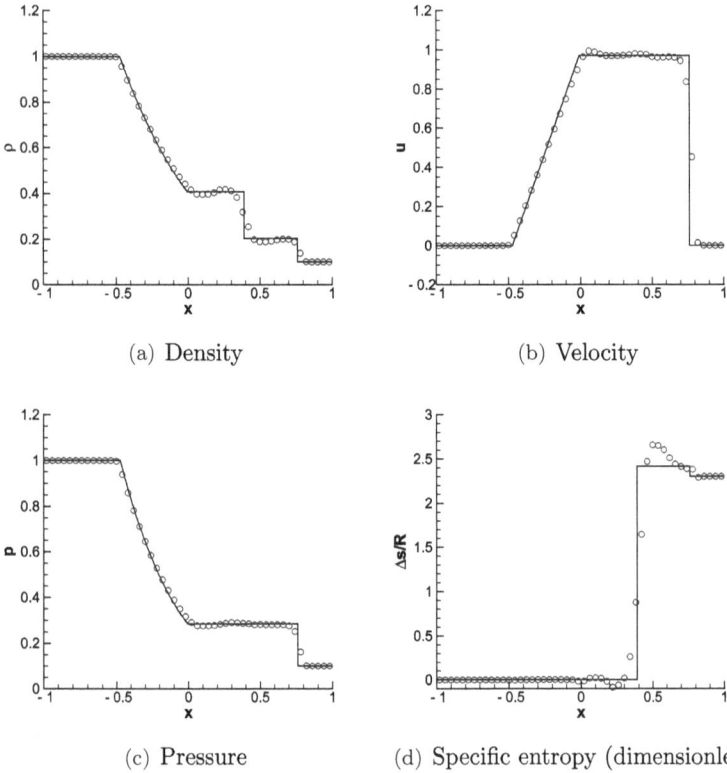

Figure 11.7:　Sod's shock tube. Comparison of the exact solution (solid line) and the solution provided using superbee limiter (circles). $t = 0.4$ time units.

First term on the left-hand side:

$\int_w^e \int_t^{t+\Delta t} \frac{\partial f}{\partial t} \, dt dx = \int_w^e f|_t^{t+\Delta t} \, dx$. The term can be approximated by assuming that function f is constant within the control volume. Therefore, the approximated term reads $(f_P^{n+1} - f_P^n)\Delta x$, where superscripts n and $n+1$ refer to current and updated times, respectively. The subscript P means that the function is evaluated at the center of the control volume.

Second term on the left-hand side:

Performing the first integration yields

$\int_t^{t+\Delta t} (uf|_e - uf|_w) \, dt$. The term can be approximated so that the value of dependent variable is constant within the volume, i.e., in Δx

and in Δt. For this term, the value of uf is assumed constant within the time interval $(t; t + \Delta t)$. Hence, the term reduces to $(uf|_e - uf|_w)\Delta t$. Notwithstanding, we do not have the values of uf at the interfaces. We will discuss this issue later on.

First term on the right-hand side:
Perform the first integration:

$\int_t^{t+\Delta t} \lambda \frac{\partial f}{\partial x}\Big|_w^e dt$. Again, assuming constant within the time span of Δt, the term becomes $\left(\lambda_e \frac{\partial f}{\partial x}\Big|_e - \lambda_w \frac{\partial f}{\partial x}\Big|_w \right) \Delta t$.

Last term on the right-hand side:
Assume that the source term is constant within the control volume. Hence, the last term can be approximated as $S_P \Delta x \Delta t$, where S is evaluated at the center of the control volume.

Hence, collecting the terms, the FV approximation of the governing equation reads

$$\left(f_P^{n+1} - f_P^n\right)\Delta x + \left(uf|_e - uf|_w\right)\Delta t = \left(\lambda_e \frac{\partial f}{\partial x}\Big|_e - \lambda|_w \frac{\partial f}{\partial x}\Big|_w \right)\Delta t$$
$$+ S_P \Delta t \Delta x. \qquad (11.105)$$

Usually, the term $\frac{\partial f}{\partial x}\Big|_e$ is approximated by $\frac{f_E - f_P}{\delta x_e}$, which is second-order accurate. Similarly, the term $\frac{\partial f}{\partial x}\Big|_w$ is approximated by $\frac{f_P - f_W}{\delta x_w}$.

To ensure the flux continuity, harmonic mean value is used for evaluating the transport property, λ. To illustrate the concept of this kind of mean value, let us take an example of heat transfer through a composite wall. A wall is made of two different materials with different thermal conductivities, say k_1 and k_2. We have to apply the concept of thermal resistance to calculate the effective thermal conductivity, see Fig. 11.8. On the one hand, the total resistance is equal to the sum of individual resistances in series, i.e.,

$$R_t = R_1 + R_2 = \frac{\Delta x_1}{2k_1} + \frac{\Delta x_2}{2k_2} = \frac{k_2 \Delta x_1 + k_1 \Delta x_2}{2k_1 k_2}. \qquad (11.106)$$

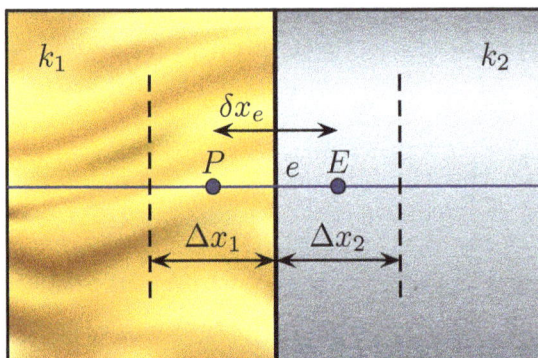

Figure 11.8:   Diagram of two different media with notations.

On the other hand, the total resistance, R_t, is equal to $\frac{\delta x}{k_f}$ or $\frac{\Delta x_1 + \Delta x_2}{2k_f}$, where k_f is the effective thermal conductivity at the interface. Therefore,

$$\frac{\Delta x_1 + \Delta x_2}{2k_f} = \frac{k_2 \Delta x_1 + k_1 \Delta x_2}{2k_1 k_2}. \tag{11.107}$$

Hence,

$$k_f = \frac{k_1 k_2 (\Delta x_1 + \Delta x_2)}{k_2 \Delta x_1 + k_1 \Delta x_2}. \tag{11.108}$$

Note: In order to ensure the flux continuity, any transport thermophysical property, such as thermal thermal diffusivity, viscosity, and mass diffusivity, must be averaged like the above equation at the interface. However, additive quantity (per unit volume), such as density and specific heat, need to be evaluated at the center of the control volume. So, there is no need for special averaging.

Approximation of the advection term: The terms $uf|_e$ and $uf|_w$ need special treatment to avoid instability issues and to maintain correct physics. If the cell Peclet number, $\text{Pe} = {}^{u\Delta x}/_{2\lambda}$, is less than one, it is a good idea to approximate the interface value as an average of the centers of the two neighboring cells. However, if the cell Peclet number is greater than one, we need to take into the consideration the direction of u. If $u > 0$, the West interface value should be biased toward the West cell, i.e., $uf|_w = u_W f_W$ and East interface value should be biased toward the cell P on which the integral is performed, i.e., $uf|_e = u_P f_P$. This approach is called **upwind scheme**. Of course, if $u < 0$, then $uf|_e = u_E f_E$ and $uf|_w = u_P f_P$. However, the upwind scheme is first-order accurate, which is not desirable. A more common scheme is the quadratic upwind interpolation for convective kinematics (**QUICK**) scheme, which is second-order accurate. The interface value is evaluated using quadratic polynomial considering the flow direction. For example, if the flow is from left to right ($u > 0$), then the dependent variable value at the East interface should consider the values of the dependent variable at cells W, P, and E. If the flow is from right to left ($u < 0$), then the value at the East interface includes the contribution of cells EE, E, and P.

The final FV scheme for the advection-diffusion equation for the one-dimensional problem (11.103) can be written in general form as

$$a_p \phi_p = a_E \phi_E + a_W \phi_w + S_p, \qquad (11.109)$$

where $a_p = a_E + a_W$ and S_p is the source term that includes the contribution of the current time solution among others.

11.9 Examples

Example 1: In many manufacturing processes, a material (stainless steel, plastics, etc.) is withdrawn from furnaces at a certain speed for cooling purposes. Consider a stainless-steel sheet coming out of a furnace at $500°$C to a room temperature of $30°$C. The steel sheet thickness is 2.5 cm and is 100 cm long. The steel sheet leaves the furnace at a speed of 0.1 m/s. Estimate the temperature distribution along the steel sheet when the heat exchange coefficient to the room is 200 W/m^{2}K. The thermal conductivity, density, and specific heat of the steel are 60 W/m K, 8000 kg/m^3, and 850 J/kg K, respectively. Compare the predicted results using finite difference and finite volume methods with the analytical solution.

Solution: In steady state, the governing equation without radiation effects is

$$\rho A_c c_p u \frac{dT}{dx} = A_c k \frac{d^2 T}{dx^2} + h P (T - T_r), \qquad (11.110)$$

where A_c is the cross-sectional area of the steel sheet, which is equal to the thickness multiplied by the width of the steel sheet (tW). c_p, ρ, and k are the specific heat, density, and thermal conductivity of the steel, respectively. P is the perimeter ($2W + 2t \approx 2W$). h is the heat exchange coefficient. The dimensionless governing equation is formulated as

$$\frac{d\phi}{dx^*} = \frac{1}{Pe}\frac{d^2\phi}{dx^{*2}} + \frac{BiS}{Pe}\phi. \tag{11.111}$$

The parameters are Peclet number ($Pe = \frac{u_0 L}{\alpha}$), Biot number ($Bi = \frac{hL}{k}$), and the aspect ratio ($S = L/t$). The dimensionless temperature is $\phi = \frac{T - T_r}{T_0 - T_r}$, where T_0 is the inlet temperature ($x = x^* = x/L = 0$). The boundary conditions are as follows: At $x^* = 0$, $\phi = 1$ and at $x^* = 1$, assume $\frac{d\phi}{dx^*} = 0$. Use $\Delta x^* = 0.05$ and 0.01.

Example 2: The first example was an ordinary equation. However, the concept is the same for partial differential equations. Steady boundary layer equation for a flow over a flat plate can be formulated as follows:

Momentum:

$$\frac{\partial u}{\partial t} + u\frac{\partial u}{\partial x} + v\frac{\partial u}{\partial y} = \nu\frac{\partial^2 u}{\partial y^2}, \tag{11.112}$$

Continuity:

$$\frac{\partial u}{\partial x} + \frac{\partial v}{\partial y} = 0, \tag{11.113}$$

with the following boundary conditions: At $x = 0$, $u = U_0$ and $v = 0$; at $y = 0$, $u = v = 0$; at $y = \infty$, $u = U_0$ and $v = 0$. Although considered in steady state, showing the parabolic character of the problem is straightforward, the $x-$ variable is playing the role of the "time-like" variable. It follows that there is no need to prescribe any downstream boundary condition in order for the problem to be solved. So, the solution strategy is to march along $x-$ direction while solving along $y-$ direction.

In dimensionless form, the momentum equation is as follows

$$\frac{\partial U}{\partial \tau} + U\frac{\partial U}{\partial X} + V\frac{\partial U}{\partial Y} = \frac{1}{Re}\frac{\partial^2 U}{\partial Y^2}, \qquad (11.114)$$

while the continuity equation reads

$$\frac{\partial U}{\partial X} + \frac{\partial V}{\partial Y} = 0, \qquad (11.115)$$

with the boundary conditions as follows: At $X = 0$, $U = 1$ and $V = 0$; at $Y = 0$, $U = V = 0$; at $Y = \infty$, $U = 1$ and $V = 0$.

Solve the above equation for Re $= 100$, $Y \in (0, 1]$, $X \in (0, 10]$, by using $\Delta Y = 0.02$ and $\Delta X = 0.5$. Use finite difference method with upwind scheme for the advection term. March the solution with the time-like variable x until the change in the velocity field is very small (in the order of 10^{-3}). Select $\Delta \tau$ considering the stability condition.

11.10   Vorticity–Stream Function Formulation

The two-dimensional Navier–Stokes equations of an incompressible fluid flow can be formulated in terms of vorticity transport equations as

$$\frac{\partial \omega}{\partial \tau} + U\frac{\partial \omega}{\partial X} + V\frac{\partial \omega}{\partial Y} = \frac{1}{Re}\left(\frac{\partial^2 \omega}{\partial X^2} + \frac{\partial^2 \omega}{\partial Y^2}\right). \qquad (11.116)$$

where ω is the vorticity, U and V the velocity components, and Re the Reynolds number. The stream function, ψ, satisfies the following Poisson equation:

$$\frac{\partial^2 \psi}{\partial X^2} + \frac{\partial^2 \psi}{\partial Y^2} = -\omega. \qquad (11.117)$$

The stream function is linked with the velocity components as follows:

$$U = \frac{\partial \psi}{\partial Y} \qquad (11.118)$$

and

$$V = -\frac{\partial \psi}{\partial X}. \qquad (11.119)$$

Note that because the fluid flow problem considered here is two-dimensional, both the vorticity and the stream function are scalars. For three-dimensional fluid flow problems, they have to be viewed as three-component (in general) vectors.

This is a typical example where a parabolic advection–diffusion equation is strongly coupled with an elliptic Poisson equation. The main task in such a formulation lies in translating boundary conditions expressed in primitive variables (velocity and pressure) into boundary conditions expressed in the derived variables (vorticity–stream function).

Problem 1: Predict the flow field in a closed cavity, with a moving lid that has dimensionless velocity of $U_{\text{lid}} = 1$. The cavity has dimensionless length of $L = 1$ and dimensionless height of $H = 1$, see Fig. 11.9.

Use finite difference method with upwind scheme for the advection terms. The plot of streamlines for $\text{Re} = U_{\text{lid}}L/\nu = 100$, where ν is the kinematic viscosity of the fluid, is shown in Fig. 11.10.

Boundary conditions in primitive variables are as follows:

- At the boundaries of the cavity, except at the lid, $U = V = 0$.
- At the lid, $U = 1$ and $V = 0$.

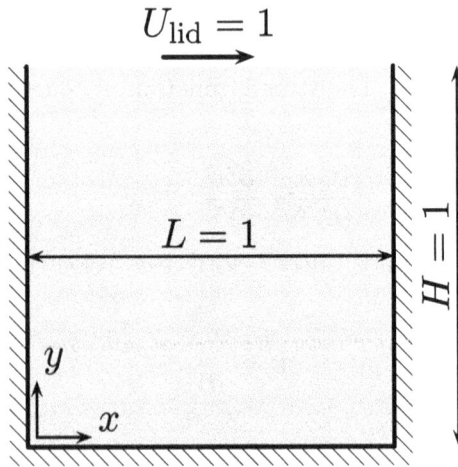

Figure 11.9:   Lid-driven cavity problem. Hatched boundaries are fixed walls.

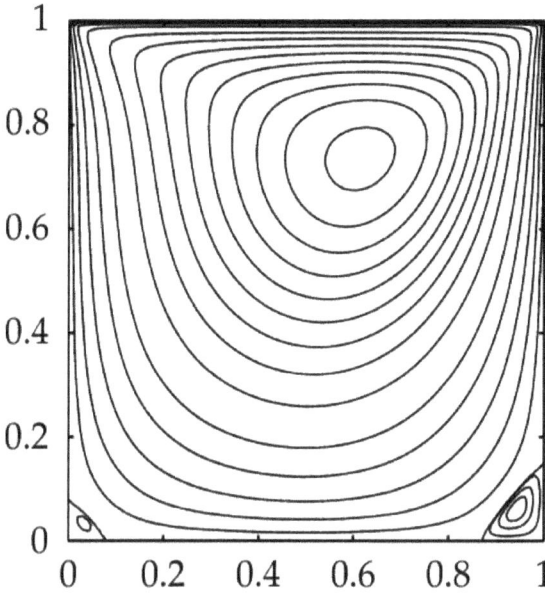

Figure 11.10:   Streamlines, lid-driven cavity.

Boundary conditions in ψ variable are as follows:

- At the left and right walls, $\psi = \frac{\partial \psi}{\partial X} = 0$ and at $Y = 0$, $\psi = \frac{\partial \psi}{\partial Y} = 0$.
- At the lid, $Y = 1$, $\psi = 0$, $\frac{\partial \psi}{\partial Y} = 1$.

To set the boundary conditions for ω, use Poisson equation, i.e.,

- for $X = 0$ and $X = 1$, $\omega = -\frac{\partial^2 \psi}{\partial X^2}$.
- for $Y = 0$ and $Y = 1$, $\omega = -\frac{\partial^2 \psi}{\partial Y^2}$.

Problem 2: Two species of concentrations α and β are reacting in a porous medium with a main flow velocity $\mathbf{U}$. The equations governing the problem read

$$
\begin{aligned}
\frac{\partial \alpha}{\partial t} + \mathbf{U} \cdot \nabla \alpha &= \kappa \left(\beta - \alpha \right) + \nabla \cdot (D_\alpha \nabla \alpha), \\
\frac{\partial \beta}{\partial t} + \mathbf{U} \cdot \nabla \beta &= \kappa \left(\alpha - \beta \right) + \nabla \cdot (D_\beta \nabla \beta),
\end{aligned}
\tag{11.120}
$$

where κ is the rate constant of the reaction and D_α and D_β the diffusivities of species α and β, respectively. This is a simplified model of a COVID-19 antigen test kit. For a one-dimensional problem inside a channel of length $L = 4 \times 10^{-2}$ m, the following initial and boundary conditions are considered:

$$\begin{cases} \alpha(x, t = 0) = 0, \\ \beta(x, t = 0) = 10^{-1} \end{cases} \qquad (11.121)$$

(both in mole/m^3) and

$$\begin{cases} \alpha = 2 \times 10^{-1}, \quad \dfrac{\partial \beta}{\partial x} = 0 \quad \text{for } x = 0 \quad \text{(i.e., inlet)}, \\ \dfrac{\partial \alpha}{\partial x} = 0, \qquad \dfrac{\partial \beta}{\partial x} = 0 \quad \text{for } x = L \quad \text{(i.e., outlet)}. \end{cases} \qquad (11.122)$$

Setting $U = 5 \times 10^{-4}$ m/s, $\kappa = 10^{-3}$ s^{-1}, and $D_\alpha = D_\beta = 10^{-5}$ m^2/s, solve the advection-diffusion problem using the finite volume method in the time interval $[0, 100]$ (in s). Plot α versus time for $x = L$ and deduce the time constant, τ, of the kit. (Recall that τ is such that $\frac{\alpha(L, \tau)}{\alpha_{\text{steady}}} = 1 - e^{-1}$, where α_{steady} is the steady-state concentration.)

Extra Reading

R. H. Sanders and K. H. Prendergast. The possible relation of the 3-kiloparsec arm to explosions in the galactic nucleus. *The Astrophysical Journal*, 188:489, 1974.

J. L. Steger and R. F. Warming. Flux vector splitting of the inviscid gas-dynamic equations with application to finite-difference methods. *Journal of Computational Physics*, 40(2):263–293, 1981. doi: https://doi.org/10.1016/0021-9991(81)90210-2. https://www.sciencedirect.com/science/article/pii/0021999181902102.

S. K. Godunov and I. Bohachevsky. A difference scheme for numerical solution of discontinuous solution of hydrodynamic equations. *Math. Sbornik*, 47:271–306, 1959. https://hal.archives-ouvertes.fr/hal-01620642/document.

P. L. Roe. Approximate Riemann solvers, parameter vectors, and difference schemes. *Journal of Computational Physics*, 43(2):357–372, 1981. doi: https://doi.org/10.1016/0021-9991(81)90128-5. https://www.sciencedirect.com/science/article/pii/0021999181901285.

Chapter 12

Wave Equation

Wave equation is a hyperbolic type of equation. Wave equation simulates many engineering problems, such as vibrations of structures (e.g., membranes, stings, and beams), sound waves, seismic waves, light waves, and electromagnetic waves (Maxwell's equation). A one-dimensional wave equation can be written in Cartesian coordinate system as

$$\frac{\partial^2 f}{\partial t^2} = c^2 \frac{\partial^2 f}{\partial x^2}, \tag{12.1}$$

where f is the dependent variable, such as pressure or displacement and c is wave speed. The equation has second derivative in time and space. Hence, there is a need for two initial and two boundary conditions.

The general form of the wave equation can be written as

$$\frac{\partial^2 f}{\partial t^2} = c^2 \nabla^2 f, \tag{12.2}$$

where ∇^2 is Laplace operator.

The wave equation can be split into two coupled first-order advection equations. For one-dimensional wave equation, the system of first-order coupled equations is

$$\frac{\partial f}{\partial t} + c \frac{\partial g}{\partial x} = 0 \tag{12.3}$$

and

$$\frac{\partial g}{\partial t} + c\frac{\partial f}{\partial x} = 0. \tag{12.4}$$

Hence, the methods discussed in Chapter 11 can be applied to solve the above system of equations as well.

The exact and general solution of the one-dimensional wave equation is

$$f(x,t) = F(x - ct) + G(x + ct), \tag{12.5}$$

where F and G are any twice differentiable functions.

For example, consider the above one-dimensional wave equation with the following boundary conditions:

$$f(0,t) = a \quad \text{and} \quad f(L,t) = b, \tag{12.6}$$

where a and b are some constants and L is the length of the domain (a string, for instance). The initial conditions are

$$f(x,0) = k(x) \quad \text{and} \quad \frac{\partial f}{\partial t}(x,0) = h(x), \tag{12.7}$$

where $k(x)$ and $h(x)$ are given functions of x. Using explicit finite difference approximation for Eq. (12.1) yields

$$\frac{f_i^{n+1} - 2f_i^n + f_i^{n-1}}{\Delta t^2} = c^2 \frac{f_{i+1}^n - 2f_i^n + f_{i-1}^n}{\Delta x^2}. \tag{12.8}$$

Rearranged, the equation reads

$$f_i^{n+1} = \lambda^2 f_{i+1}^n + 2(1 - \lambda^2)f_i^n + \lambda^2 f_{i-1}^n - f_i^{n-1}, \tag{12.9}$$

where $\lambda = {}^{c\Delta t}/_{\Delta x}$ is the usual Courant number. For the first-step solution, the first derivative can be approximated by extending the time domain artificially and using central difference (see Fig. 12.1) as

$$\frac{f_i^1 - f_i^{-1}}{2\Delta t} = h(x). \tag{12.10}$$

Hence, $f_i^{-1} = f_i^1 - 2\Delta t\, h(x)$. Therefore, for $n = 0$, the FD equation reads

$$f_i^1 = \lambda^2 f_{i+1}^0 + 2(1 - \lambda^2)f_i^0 + \lambda^2 f_{i-1}^0 - f_i^1 + 2\Delta t\, h(x). \tag{12.11}$$

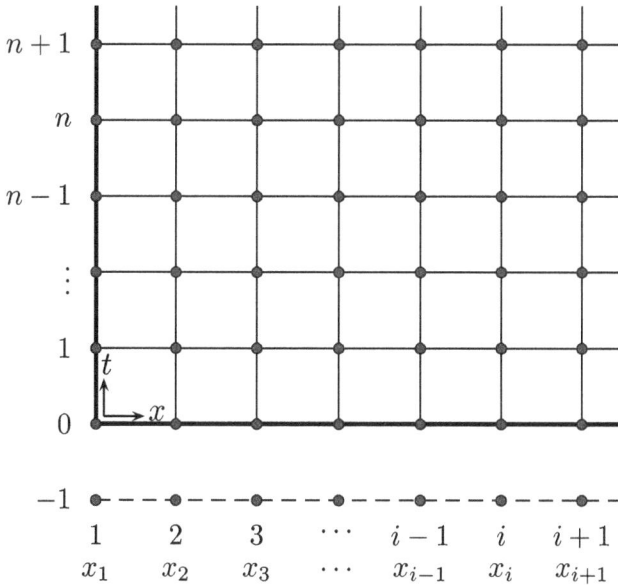

Figure 12.1: Space–time grid for solving wave equation.

The extension of the above process for two- or three-dimensional problems is straightforward.

12.1 Consistency

Let us use Taylor series expansion for f_i^{n+1}, f_{i+1}^n, f_{i-1}^n, and f_i^{n-1} (index i is omitted in derivatives for brevity):

$$f_i^{n+1} - f_i^n + \frac{\partial f}{\partial t}\Delta t + \frac{1}{2!}\frac{\partial^2 f}{\partial t^2}\Delta t^2 + \frac{1}{3!}\frac{\partial^3 f}{\partial t^3}\Delta t^3 + O(\Delta t^4), \quad (12.12)$$

$$f_{i+1}^n = f_i^n + \frac{\partial f}{\partial x}\Delta x + \frac{1}{2}\frac{\partial^2 f}{\partial x^2}\Delta x^2 + O(\Delta x^3), \quad (12.13)$$

$$f_{i-1}^n = f_i^n - \frac{\partial f}{\partial x}\Delta x + \frac{1}{2}\frac{\partial^2 f}{\partial x^2}\Delta x^2 + O(\Delta x^3), \quad (12.14)$$

and

$$f_i^{n-1} = f_i^n - \frac{\partial f}{\partial t}\Delta t + \frac{1}{2}\frac{\partial^2 f}{\partial t^2}\Delta t^2 - \frac{1}{3!}\frac{\partial^3 f}{\partial t^3}\Delta t^3 + O(\Delta t^4). \quad (12.15)$$

Substituting into Eq. (12.9) yields

$$\frac{\partial^2 f}{\partial t^2} = c^2 \frac{\partial^2 f}{\partial x^2} + O(\Delta x^2) + O(\Delta t^2). \qquad (12.16)$$

Hence, the scheme is consistent and is second order in space and time locally.

12.2   Stability Analysis

The function f is expanded using Fourier series, $f_i^n = MG^n e^{Iki\Delta x}$. Hence,

$$f_i^{n+1} = MG^{n+1} e^{Iki\Delta x}$$

$$= MG^n e^{Iki\Delta x} \left(\lambda^2 e^{Ik\Delta x} + 2(1 - \lambda^2) + \lambda^2 e^{-I\Delta kx} \right)$$

$$- MG^{n-1} e^{Iik\Delta x}. \qquad (12.17)$$

The term $e^{Iik\Delta x\theta}$ can be canceled. Also, $\cos\theta = \frac{e^{I\theta} + e^{-I\theta}}{2}$. Hence,

$$G^{n+1} = 2(1 + \lambda^2(\cos\theta - 1))G^n - G^{n-1}, \qquad (12.18)$$

which can be rewritten as

$$G^{n+1} = 2\left(1 - 2\lambda^2 \sin^2 \left(\frac{\theta}{2} \right) \right) G^n - G^{n-1}. \qquad (12.19)$$

Dividing this equation by G^n gives

$$G = 2\left(1 - 2\lambda^2 \sin \left(\frac{\theta}{2} \right) \right) - \frac{1}{G}, \qquad (12.20)$$

which leads to

$$G^2 - 2\gamma G + 1 = 0, \qquad (12.21)$$

where $\gamma = 1 - 2\lambda^2 \sin\left(\frac{\theta}{2}\right)$. The equation is a quadratic algebraic equation. Its solutions are

$$G_{1,2} = \gamma \pm \sqrt{\gamma^2 - 1}. \qquad (12.22)$$

In order for the scheme to be stable, Courant–Friedrichs–Lewy (CFL) number should be less than or equal to unity, i.e., $\frac{c\Delta t}{\Delta x} \leqslant 1$. (The details of the derivation are left as an exercise).

12.3 Worked Example

Let us use the FDM to solve the following equation:

$$\frac{\partial^2 f}{\partial t^2} = c^2 \frac{\partial^2 f}{\partial x^2} \quad \text{for } 0 < x < 1, \tag{12.23}$$

which is subjected to the following boundary and initial conditions:

$$f(t, 0) = 0, \qquad f(t, 1) = 0,$$

$$f(0, x) = \sin(\pi x), \quad \frac{\partial f}{\partial t}(0, x) = 0. \tag{12.24}$$

We can compare the numerical predictions with the analytical solution, $f_{\text{exact}}(t, x) = \cos(c\pi t)\sin(\pi x)$. Set $c = 5$ and use $\Delta x = 0.1$ and $\Delta x = 0.2$. To satisfy the stability condition ($\frac{c\Delta t}{\Delta x} \leqslant 1$), we set Δt such that $\lambda = \frac{c\Delta t}{\Delta x} = 0.5$ and carry out the computations for the total time of $t^{\text{final}} = 10$.

Solution: Let us denote by f_i^n the approximation of the solution at (x_i, t^n), and let N be the total number of space nodes used. So, $\Delta x = \frac{1}{N-1}$ and $x_i = \Delta x(i-1)$, $i = 1, \ldots, N$.

According to the boundary conditions, we have

$$\begin{cases} f_1^n = 0 \\ f_N^n = 0 \end{cases} \quad \text{for all } n. \tag{12.25}$$

According to the initial conditions,

$$f_i^0 = f(x_i, 0) = \sin(\pi x_i),$$

and setting $h(x) = 0$ in Eq. (12.11), we get

$$f_i^1 = \frac{1}{2}\left(\lambda^2 f_{i+1}^0 + 2(1 - \lambda^2)f_i^0 + \lambda^2 f_{i-1}^0\right), \quad i = 2, \ldots, N - 1. \tag{12.26}$$

The latter equation is valid only at the first computed time step. For the other time steps, Eq. (12.9), for $i = 2, \ldots, N - 1$, is used in conjunction with boundary conditions (12.25).

It is worth noting that using a uniform mesh, the one-dimensional FVM would lead to exactly the same Eq. (12.9). Different results are

Table 12.1: FDM using two different space steps: Analytical tabulated solutions compared with the analytical solution.

x_i	FDM with $N = 6$	FDM with $N = 11$	Analytical
0.0000000000000000	0.0000000000000000	0.0000000000000000	0.0000000000000000
0.1000000000000000		0.2938926261462330	0.2938926261464249
0.2000000000000000	0.1406546275136255	0.5590169943749474	0.5590169943753057
0.3000000000000000		0.7694208842938123	0.7694208842943065
0.4000000000000000	0.2275839679920023	0.9045084971874737	0.9045084971880533
0.5000000000000000		0.9510565162951526	0.9510565162957631
0.6000000000000000	0.2275839679920007	0.9045084971874737	0.9045084971880533
0.7000000000000000		0.7694208842938136	0.7694208842943065
0.8000000000000000	0.1406546275136218	0.5590169943749474	0.5590169943753058
0.9000000000000000		0.2938926261462403	0.2938926261464250
1.0000000000000000	0.0000000000000000	0.0000000000000000	1.164669852699966E-016

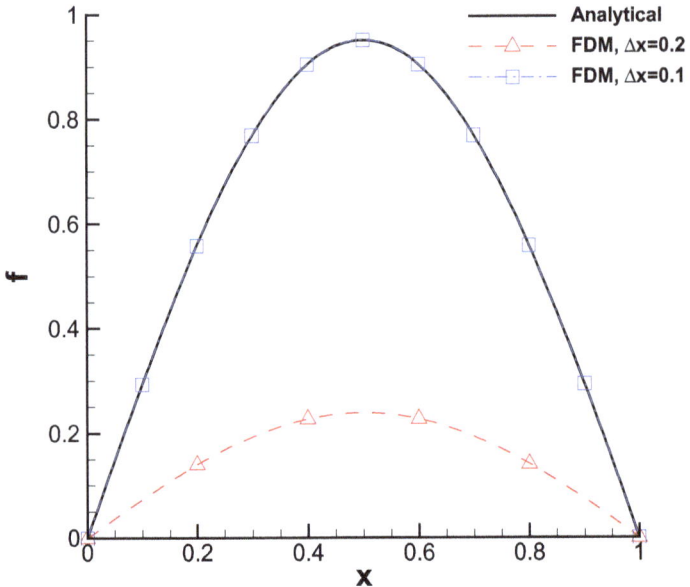

Figure 12.2: FDM solution using two different space steps compared with the analytical solution.

expected when the problem is two-dimensional (or higher) or when the mesh used is not uniform.

The solution at time $t = 10$ is computed using $\Delta x = 0.1$ ($N = 11$) and $\Delta x = 0.2$ ($N = 6$) and compared with the analytical solution in Table 12.1 and plotted in Fig. 12.2.

For the larger space step, $\Delta x = 0.2$, the solution is far from the analytical solution: The numerical solution is in this case highly diffused. For the smaller space step, $\Delta x = 0.1$, the difference between the numerical solution and the analytical one is graphically imperceptible. Actually, inspecting Table 12.1, the mean difference, $\frac{1}{N} \sum_{i=1}^{N} |f_i^{n_{\text{final}}} - f_{\text{exact}}(t^{n_{\text{final}}}, x_i)|$, calculated is less than 10^{-12} in this case. In other words, the second-order FDM method used for this specific problem is impressively accurate.

Note: It is always preferable to non-dimensionalize the problem. For the above problem, it is better to introduce dimensionless length defined as $\hat{x} = x/L$, where L is the domain length. Also, dimensionless time can be expressed as $\tau = tL/c$. The final equation reads

$$\frac{\partial^2 f}{\partial \tau^2} = \frac{\partial^2 f}{\partial \hat{x}^2}, \tag{12.27}$$

with boundary conditions of $f(\tau, 0) = 0$ and $f(\tau, 1) = 0$ and initial conditions of $f(0, \hat{x}) = \sin(\pi \hat{x})$ and $f_t(0, \hat{x}) = 0$. In the following, we elaborate on one of the methods discussed in Chapter 11.

12.4   Lax–Wendroff Method

The Lax–Wendroff method (LWM) is explicit, second-order accurate in space and time. For a one-dimensional problem, using Taylor series for f_i^{n+1} yields (subscript i and superscript n are omitted in derivatives for brevity)

$$f_i^{n+1} = f_i^n + \frac{\partial f}{\partial t} \Delta t + \frac{1}{2!} \frac{\partial^2 f}{\partial t^2} \Delta t^2 + O(\Delta t)^3. \tag{12.28}$$

From Eq. (12.3), $\frac{\partial f}{\partial t} = -c \frac{\partial g}{\partial x}$. So, substituting in the above equation gives

$$f_i^{n+1} = f_i^n - c \frac{\partial g}{\partial x} \Delta t - \frac{c}{2!} \frac{\partial^2 g}{\partial t \partial x} \Delta t^2 + O(\Delta t)^3. \tag{12.29}$$

Furthermore, differentiating Eq. (12.3) with respect to time,

$$\frac{\partial^2 f}{\partial t^2} = -c \frac{\partial}{\partial x} \left(-c \frac{\partial f}{\partial x} \right) \tag{12.30}$$

and using Eq. (12.4), Eq. (12.29) becomes

$$f_i^{n+1} = f_i^n - c\frac{\partial g}{\partial x}\Delta t + \frac{c^2}{2}\frac{\partial^2 f}{\partial x^2}\Delta t^2 + O(\Delta t)^3. \tag{12.31}$$

Using central difference for space approximation for both derivatives involved in the above equation, we get

$$f_i^{n+1} = f_i^n - \frac{c\Delta t}{2\Delta x}(g_{i+1}^n - g_{i-1}^n) + \frac{c^2\Delta t^2}{2\Delta x^2}(f_{i+1}^n - 2f_i + f_{i-1}^n). \tag{12.32}$$

Similarly, g_i^{n+1} can be approximated as

$$g_i^{n+1} = g_i^n - \frac{c\Delta t}{2\Delta x}(f_{i+1}^n - f_{i-1}^n) + \frac{c^2\Delta t^2}{2\Delta x^2}(g_{i+1}^n - 2g_i^n + g_{i-1}^n). \tag{12.33}$$

Both f and g can be updated in time using initial values of the variables. In fact, initial conditions (12.7) read in terms of f and g variables as

$$f(x,0) = k(x) \quad \text{and} \quad g(x,0) = \frac{1}{c}\int_0^x h(\xi)d\xi. \tag{12.34}$$

(Note that g in Problem (12.3) and (12.4) is defined within an additive constant. So, setting $g(0,0) = 0$ does not affect the solution f.) Similarly, we can apply flux vector splitting methods mentioned in Chapter 11.

12.5 Implicit Method

Explicit methods are conditionally stable, which restricts the time step value. Implicit methods are (heuristically) unconditionally stable. However, they require to solve a system of algebraic equations for each time step. To establish implicit method, Taylor series is applied by expanding the dependent variable and its derivative with time and space. For a two-dimensional problem, it reads

$$\frac{\partial^2 f}{\partial t^2} = c^2\left(\frac{\partial^2 f}{\partial x^2} + \frac{\partial^2 f}{\partial y^2}\right). \tag{12.35}$$

The approximations lead to

$$\frac{f_{i,j}^{n+1} - 2f_{i,j}^{n} + f_{i,j}^{n-1}}{\Delta t^2}$$

$$= c^2 \left[\frac{f_{i+1,j}^{n+1} - 2f_{i,j}^{n+1} + f_{i-1,j}^{n+1}}{\Delta x^2} + \frac{f_{i,j+1}^{n+1} - 2f_{i,j}^{n+1} + f_{i,j-1}^{n+1}}{\Delta y^2} \right].$$

$$(12.36)$$

The system of algebraic equations can be solved by methods introduced in Chapter 2 at each time step.

12.6   Problems

Project 1: A typical cloud-to-ground lightning is a series (of ten, usually) of flashes, each of them lasting about 30×10^{-6} s. We intend to calculate the magnetic field potential that one flash induces. The discharge core is a cylindrical shape of radius $r_0 = 10^{-2}$ m roughly. The current density versus time can be approximated using the following law:

$$j(t,r) = \begin{cases} \dfrac{I}{\pi r_0^2 \left(\dfrac{t}{a}\right)^3 \left(\exp\left(\dfrac{a}{t}\right) - 1\right)} & \text{for } 0 \leqslant r \leqslant r_0, \\ 0 & \text{for } r > r_0, \end{cases} \quad (12.37)$$

where $a = 3$ μs and $I = 3 \times 10^4$ A. Note that outside the lightning, i.e., for $r > r_0$, the current density is set to zero.

The magnetic potential A obeys the following axisymmetric law deduced from Maxwell's equations:

$$\frac{\partial^2 A}{\partial t^2} - c^2 \frac{1}{r} \frac{\partial}{\partial r} \left(r \frac{\partial A}{\partial r} \right) = \mu_0 c^2 j, \quad (12.38)$$

where c is the speed of light ($c = 3 \times 10^8$ m/s) and $\mu_0 = 4\pi \times 10^{-7}$ H/m is the vacuum permeability. The boundary conditions are

$$\begin{cases} \dfrac{\partial A}{\partial r} = 0 & \text{for } r = 0, \\ \dfrac{\partial A}{\partial t} - c \dfrac{\partial A}{\partial r} = 0 & \text{for } r = L, \end{cases} \quad (12.39)$$

where $L = 10$ m, and the initial conditions are

$$
\begin{cases}
A = 0 \\
\dfrac{\partial A}{\partial t} = 0
\end{cases}
\quad \text{for } t = 0.
\tag{12.40}
$$

Solve this wave problem using FDM and FVM up to $t = 30 \times 10^{-6}$ s. (Use implicit Euler scheme for both methods in order to ameliorate the time step and recall that the indetermination of the term $\frac{1}{r}\frac{dA}{dr}\big|_{r=0}$ can be resolved by using L'Hospital's rule as presented in Chapters 7 and 9.) Plot, versus time, the magnetic field magnitude at $r = L$ using the following formula: $B = \left|\frac{\partial A}{\partial r}\right|$.

Project 2: The equation that describes the light propagation in graded-index optical fibers reads

$$
\frac{\partial^2 A}{\partial t^2} - \frac{c^2}{\left(n_1 - \left(\frac{r}{r_0}\right)^2 (n_1 - n_2)\right)^2}
$$

$$
\times \left(\frac{1}{r}\frac{\partial}{\partial r}\left(r\frac{\partial A}{\partial r}\right) + \frac{1}{r^2}\frac{\partial^2 A}{\partial \theta^2} + \frac{\partial^2 A}{\partial z^2}\right) = 0,
\tag{12.41}
$$

where t, r, θ, and z are the cylindrical coordinates and A the magnetic potential. c the speed of light, r_0 the fiber core radius, and n_1 and n_2 the refractive index limits of the fiber, $n_1 > n_2$, and $0 \le r \le r_0$.

Setting $c = 3 \times 10^8$ m/s, $n_1 = 1.5$, and $n_2 = 1.45$, $r_0 = 2.5$ μm, and considering the following boundary conditions:

$$
\begin{cases}
A = 0 & \text{for } r = r_0, \\
A = D\cos(2\pi f t) & \text{for } z = 0, \\
\dfrac{\partial A}{\partial t} - c\dfrac{\partial A}{\partial z} = 0 & \text{for } z = \lambda
\end{cases}
\tag{12.42}
$$

$$
\left(D = \left[\frac{\cos(\pi r^2/2r_0^2)}{\cos\pi/4}\right]\exp\left(-20\left[\left(\frac{r}{r_0}\cos\theta - \frac{1}{2}\right)^2 + \left(\frac{r}{r_0}\sin\theta\right)^2\right]\right)\right)
$$

and the following initial conditions:

$$
\begin{cases}
A = 0 \\
\dfrac{\partial A}{\partial t} = 0
\end{cases}
\quad \text{for } t = 0,
\tag{12.43}
$$

solve the three-dimensional problem (in cylindrical coordinates) for a distance $\lambda = 0.7 \times 10^{-6}$ m and a frequency $f = c/\lambda$ using the explicit Euler scheme and the FDM.

Project 3: The linearized shallow-water equation reads

$$\frac{\partial^2 \left(\sqrt{g}\eta\right)}{\partial t^2} - \nabla \cdot [g\, h \cdot \nabla \left(\sqrt{g}\eta\right)] = 0, \qquad (12.44)$$

where η is the surface height from the geoid, see Fig. 12.3, h is the seabed topology, and g is the acceleration of gravity. This equation describes roughly the tsunami phenomenon (water waves caused by earthquakes, volcanic eruptions, etc.). We assume that the problem is one-dimensional, and the seabed described by the following equation:

$$h(x) = \exp\left(-\frac{x}{2}\right) - \frac{\sin(2\pi x)}{10x}, \qquad (12.45)$$

with the following initial conditions:

$$\begin{cases} \eta(x, t = 0) = a \exp(-5x), \\ \dfrac{\partial \eta}{\partial t} = 0 \end{cases} \qquad (12.46)$$

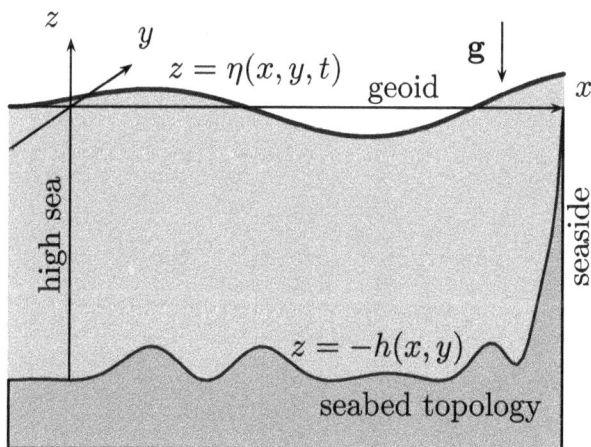

Figure 12.3: Shallow-water problem.

and the following boundary conditions:

$$
\begin{cases}
\dfrac{\partial \eta}{\partial x} = 0 & \text{for } x = 0 \quad \text{(i.e., high sea)}, \\[2ex]
\dfrac{\partial \eta}{\partial t} - g\, h(x) \dfrac{\partial \eta}{\partial x} = 0 & \text{for } x = 2\pi \quad \text{(i.e., seaside)},
\end{cases}
\tag{12.47}
$$

where a is proportional to the magnitude of the energy causing the initial wave. By taking $a = 0.01$, 0.2, 0.5, and 1, solve the problem between $0 \leqslant x \leqslant 2\pi$ using the FVM. Plot at seaside, i.e., for $x = 2\pi$, the maximum height of water versus $\ln a$.

Index

A

adjoined problem, 149
advection equation, 221
alternating direction implicit (ADI),
    187, 215
amplification factor, 175, 178, 185
approximations, 93

B

backward, 93
backward expansion, 10
bi-conjugate gradient, 43
bisection method, 76
boundary value problems, 145, 149,
    151, 154

C

central difference, 93
chain method, 216
Chebyshev method, 58, 81
Chebyshev points, 62
Chebyshev polynomial, 58
Choleski, 33
condition number, 23
conjugate gradient, 43
conjugate gradient method, 48
consistency, 96, 133
consistency analysis, 133, 172, 186
convergence, 136, 171
Cramer's rule, 26

Crout, 33
cubic splines, 67, 69–70

D

determinant, 21
diagonal dominance, 23
diagonal elements, 20
diagonally dominant, 34
direct methods, 26
discretization errors, 5, 8
Doolittle, 33
dot product, 18

E

eigenvalue, 22, 25, 48
elliptic, 169–170, 205, 215, 262
Euclidean, 19
Euclidean norm, 22
Euler method, 122–124, 127, 132,
    134–136, 138–142
existence, 172
extrapolation, 113

F

finite volume method (FVM), 154,
    159, 161, 189, 214, 231, 248, 254,
    264
fixed point method, 77
forward, 93
forward expansion, 10

G

Gauss elimination, 27, 30
Gauss quadrature, 109
Gauss–Jordan method, 31
Gauss-Seidel method, 35, 38, 43

H

Hermite, 55
higher-order approximations, 95
higher-order polynomial, 55
hyperbolic, 169–170, 174, 221, 226,
    232–233, 236, 238, 265

I

identity matrix, 20
ill-conditioned matrix, 23
initial value problems, 117–118,
    145–146, 182
inverse, 21
iterative methods, 26, 34

J

Jacobi method, 35, 37–38, 43

L

Lagrange, 55
Lagrange interpolation, 55
Lagrange polynomial, 62
Laplace equation, 33, 172, 205–206,
    218–219
Lax–Milgram theorem, 171
Lax–Richtmyer theorem, 170–171
leapfrog method, 180, 225
least square method, 64
linear algebra, 17, 19
linear interpolation, 54
linear shooting method, 146,
    166
lower triangular matrix, 20
LU factorization, 31

M

marching-like methods, 149, 174
matrix, 19

method of lines, 149, 182
Monte Carlo method, 112
multi-variable fitting, 65
multi-variable functions, 13

N

Newton–Raphson method, 79–81,
    148, 161
non-dimensional, 5
non-stationary methods, 34, 43
non-uniform grids, 102–103
nonlinear equation, 75, 80
nonlinear shooting method, 147,
    160–161
numerical integration, 106, 108–109

P

parabolic, 169–170, 174, 179, 230,
    260, 262
partial differential equations,
    169–172, 174, 189, 260
pentadiagonal matrix, 33
Poisson equation, 33, 205, 214,
    261–263
positive and negative definite, 23
positive definite, 48
pre-conditioner, 49

Q

quadratic equations, 44, 75
quadratic splines, 67

R

Richardson's extrapolation, 114
round-off error, 5, 7, 26

S

secant method, 82, 148–149
semi-analytical method, 120
shooting method, 160
Simpson's method, 108–109
Simpson's scheme, 106
singular values, 22
skew-symmetric matrix, 21–22

SOR method, 79
sparse matrix, 34
spectral radius, 22
spline, 55
spline interpolation, 67
stability analysis, 135, 174–177,
   184–186, 222, 224–225, 230–231,
   268
stability criterion, 176–177, 237, 239
stationary iterative methods, 34
statistical method, 112
steepest descent method, 43, 47
successive over-relaxation, 40
symmetric, 21
symmetric matrix, 22, 33, 48
system of linear algebraic equations,
   26

T

Taylor series, 8, 56, 79, 81
Thomas algorithm, 29
trace, 20
transpose, 18, 21
trapezoidal method, 106, 109, 114
tri-diagonal matrix, 21, 29
truncation, 4
truncation errors, 8

U

uniqueness, 172
upper triangular matrix, 20

V

von Neumann analysis, 175–176, 180

www.ingramcontent.com/pod-product-compliance
Lightning Source LLC
Chambersburg PA
CBHW050544190326
41458CB00007B/1912